AP® CALCULUS AB & BC
CRASH COURSE®

Flavia Banu, M.A.
Joan Rosebush, M.Ed.

Updated by Stu Schwartz

Research & Education Association
www.rea.com

ABOUT REA

Founded in 1959, Research & Education Association (REA) is dedicated to publishing the finest and most effective educational materials—including study guides and test preps—for students of all ages.

Today, REA's wide-ranging catalog is a leading resource for students, teachers, and other professionals. Visit *www.rea.com* to see our complete catalog.

Research & Education Association
258 Prospect Plains Road
Cranbury, New Jersey 08512
Email: info@rea.com

AP® CALCULUS AB & BC CRASH COURSE,® 3rd Edition

Published 2022
Copyright © 2021 by Research & Education Association.
Prior editions copyright © 2017, 2011 by Research & Education Association. All rights reserved. No part of this book may be reproduced in any form without permission of the publisher.

Printed in the United States of America

Library of Congress Control Number 2020909994

ISBN-13: 978-0-7386-1273-7
ISBN-10: 0-7386-1273-1

 REA Crash Course® and REA® are registered trademarks of Research & Education Association.

AP® Calculus AB & BC Crash Course
TABLE of CONTENTS

ABOUT OUR BOOK

REA's *AP® Calculus AB & BC Crash Course* is designed for the last-minute studier or any student who wants a quick refresher on the AP® Calculus courses. The College Board's *AP® Calculus AB and BC: Course and Exam Description* (2020) details the knowledge that students must have to succeed on the exams. This *Crash Course* aligns all the topics in the AP® Calculus courses, both AB and BC, with the latest College Board exam framework issued in 2019.

Written by veteran AP® Calculus AB and BC test experts, our *Crash Course* gives you a concise review of the major concepts and important topics tested on the AP® Calculus AB and BC exams.

- **Part I** offers you our **Keys for Success**, so you can tackle the exams with confidence.

- **Parts II through V** are the **Content Review** sections that cover all of the topics you'll find on the exams, including functions, graphs and limits, differentiation, integrals, and sequences and series.

- **Part VI** gives you techniques for using a graphing calculator and includes AP®-style practice questions, as well as tips for succeeding on both the **multiple-choice and free-response questions**.

ABOUT OUR ONLINE PRACTICE EXAMS

How ready are you for the AP® Calculus AB and BC exams? Find out by taking REA's online practice exams (one for Calculus AB and one for Calculus BC) available at *www.rea.com/studycenter*. These tests feature automatic scoring, detailed explanations of all answers, and diagnostic score reporting that will help you identify your strengths and weaknesses so you'll be ready on exam day.

Whether you use this book throughout the school year or as a refresher in the final weeks before your exam, REA's *Crash Course* will show you how to study efficiently and strategically, so you can boost your score.

Good luck on your AP® Calculus AB and BC exams!

ABOUT OUR AUTHORS

Flavia Banu earned her B.A. and M.A. in Pure Mathematics from Queens College of the City University of New York. She was an adjunct professor at Queens College from 1994 to 2008; while there, she taught Algebra and Calculus II. Currently she teaches mathematics at Bayside High School in Bayside, New York. Ms. Banu is a Math for America Master Teacher and coaches the math team at Bayside High.

Joan Marie Rosebush is a Senior Lecturer and Director of Student Success at the University of Vermont's College of Engineering and Mathematical Sciences.

Ms. Rosebush earned her bachelor of arts degree in elementary education, with a concentration in mathematics, at the University of New York in Cortland, New York. She received her master's degree in education from Saint Michael's College, Colchester, Vermont. She went on to earn a Certificate of Advanced Graduate Study in Administration at Saint Michael's College.

Stu Schwartz taught math for 35 years in the Wissahickon School District in Ambler, Pennsylvania, specializing in teaching both AP® Calculus AB and BC as well as AP® Statistics. He also has instructed at the college level: Acadia University, Montgomery County Community College, Chestnut Hill College, and Broward College. Mr. Schwartz received the Presidential Award for Excellence in Mathematics Teaching, America's highest honor for teachers of mathematics. He was a consultant to the Math and Science Partnership of Greater Philadelphia, focusing on factors affecting success in college in STEM degrees. He is well known in the AP® community, having developed the website *www.mastermathmentor.com*, which offers a trove of free curriculum-related material for teachers of AP® Calculus, AP® Statistics, and other math courses.

ACKNOWLEDGMENTS

We would like to thank Larry B. Kling, Editorial Director, for his overall guidance; Pam Weston, Publisher, for managing the publication to completion; Heidi Gagnon for digital content preparation; Wayne Barr for project management; Fiona Hallowell for proofreading; and Jennifer Calhoun for file prep.

PART I
INTRODUCTION

Keys for Success on the AP® Calculus AB & BC Exams

The AP® Calculus exams are challenging for all students. The courses have countless formulas, diagrams, and problems to review that may seem overwhelming as you near exam day. But don't worry, this *Crash Course* focuses on the key information you *really* need to know to succeed on both exams. It takes into account the three Big Ideas underpinning the course framework for 2020–2021 and beyond:

BIG IDEA 1. Change: Using derivatives to describe rates of change of one variable with respect to another or using definite integrals to describe the net change in one variable over an interval of another allows students to understand change in a variety of contexts. It is critical that students grasp the relationship between integration and differentiation as expressed in the Fundamental Theorem of Calculus—a central idea in AP® Calculus.

BIG IDEA 2. Limits: Beginning with a discrete model and then considering the consequences of a limiting case allows us to model real-world behavior and to discover and understand important ideas, definitions, formulas, and theorems in calculus: for example, continuity, differentiation, integration, and series (BC only).

BIG IDEA 3. Analysis of Functions: Calculus allows us to analyze the behaviors of functions by relating limits to differentiation, integration, and infinite series and relating each of these concepts to the others.

During the school year, you should learn most of the material that will appear on the tests. As you go through this book, if you discover that there is something you don't understand, consult your textbook or ask your teacher for clarification.

This *Crash Course* follows a practical, time-saving approach to studying for the AP® Calculus exams. It's like taking notes on flash cards, except we've already done it for you in a streamlined outline format.

I. STRUCTURE OF THE EXAM

Both the AP® Calculus AB and BC exams have the same format, as shown in the table.

	Number of Questions/ Problems	Time (minutes)
Section I: Multiple-Choice Questions Part A: No calculator	30	60
Part B: Graphing calculator allowed	15	45
Section II: Free-Response Questions Part A: Graphing calculator allowed	2	30
Part B: No calculator	4	60
TOTAL TIME		195

Each section of the test is worth 50% of your grade. You will have 3¼ hours to complete your exam. According to the College Board, in Section II of Calculus AB and BC, if you complete Part A before your time is up, you *cannot* move on to Part B. So if you finish that section early, you will have time to check your answers. However, if you complete Part B before your 60 minutes are up, you can return to Part A, without the use of the calculator. You will have enough space to work out your problems in the exam booklet.

Multiple-Choice Questions (MCQs). The first section of both the AB and BC exams contains 45 MCQs, each with four possible answers. No calculator is permitted in Part A, but a graphing calculator can be used in Part B, although not all questions necessarily require it. Both the AB and BC exams include algebraic, logarithmic, exponential, trigonometric, and piecewise functions. Both include functions that are described analytically, graphically, or in a table.

Only the first three mathematical practices are assessed in the MCQ section with the following weighting. While problems can and usually do address multiple mathematical practices, each question focuses on one major practice.

Mathematical Practice	Weighting
Practice 1: Implementing Mathematical Processes	53%–66%
Practice 2: Connecting Representations	18%–28%
Practice 3: Justification	11%–18%

The content of the questions is typically drawn from multiple units and skills within the course. But each one has a focus. For instance, a question asking the interval where an accumulation function is increasing or decreasing may ultimately require students to draw from knowledge in AP® Calculus course units 2, 3, 5, and 6. But focusing on function analysis, the question would be classified as a unit 5 problem.

Exam Weighting for the Multiple-Choice Section of the AP® Calculus Exams

	AB	BC
Unit 1: Limits and Continuity	10%–12%	4%–7%
Unit 2: Differentiation: Definition and Basic Derivative Rules	10%–12%	4%–7%
Unit 3: Differentiation: Composite, Implicit, and Inverse Functions	9%–13%	4%–7%
Unit 4: Contextual Applications of Differentiation	10%–15%	6%–9%

(cont'd)

Exam Weighting for the Multiple-Choice Section (cont'd)

	AB	BC
Unit 5: Applying Derivatives to Analyze Functions	15%–18%	8%–11%
Unit 6: Integration and Accumulation of Change	17%–20%	17%–20%
Unit 7: Differential Equations	6%–12%	6%–9%
Unit 8: Applications of Integration	10%–15%	6%–9%
Unit 9: Parametric Equations, Polar Coordinates and Vector-Valued Functions		11%–12%
Unit 10: Infinite Sequences and Series		17%–18%

The MCQs are scored electronically. There is no penalty for incorrect answers, so even if you don't know an answer, take a guess.

Free-Response Questions (FRQs). The six FRQs on the AB and BC exams assess a variety of skills across the units of the course. However, they vary from year to year and not every concept in the course is tested by FRQs. For instance, one year, an FRQ might focus on inverse functions and the next year, there might be no questions on inverse functions (although they most likely will be tested in the MCQ section). In the BC exam, years can go by without the Lagrange remainder being tested in the BC FRQ section, and then it will be tested for two straight years. It is impossible to forecast!

There are three FRQs common to the AB and BC exams. Both include various types of functions and their representations and a roughly equal mix of procedural and conceptual tasks. Impossible to test in the MCQ section, Practice 4: Communication and Notation is stressed in the free-response section. To that end, at least two of the FRQs incorporate a real-world context into the question; students are required to answer some subparts in a way that is understandable to a person without calculus knowledge.

Mathematical Practice	AB	BC
Practice 1: Implementing Mathematical Processes	37%–55%	37%–59%
Practice 2: Connecting Representations	9%–16%	9%–16%
Practice 3: Justification	37%–55%	37%–59%
Practice 4: Communication and Notation	13%–24%	9%–20%

In the free-response section of the exams, it is important to show your work so the AP® Readers can evaluate your method of answering. You will receive partial credit as long as your methods, reasoning, and conclusions are presented clearly. You should use complete sentences when answering the questions in this portion of the exam.

For those questions requiring the use of a graphing calculator, the scorers will want to see your mathematical setup that led to the solution provided by the calculator. You should demonstrate the equation being solved, derivatives being evaluated, and so on. Your answers should be in standard mathematical notation.

If a calculation is given as a decimal approximation, it should be correct to three places following the decimal point, unless you are asked for something different in the question.

II. NOTEWORTHY REVISIONS TO THE EXAMS

L'Hôspital's rule is now part of the AB exam while the more complicated forms ($\infty - \infty$, $0 \cdot \infty$, 1^{∞}, 0^0, ∞^0) are no longer part of the BC exam. Absolute and conditional convergence of infinite series has been added to the BC exam topics, as well as the limit comparison test and the alternating series error.

III. SCORING

The scores from the two sections of the Calculus exams are combined to create a composite score.

AP® SCORE SCALE

5	Extremely well qualified	A+, A
4	Well qualified	A–, B+, B
3	Qualified	B–, C+, C
2	Possibly qualified	
1	No recommendation	

To be "qualified" is to receive college credit or advanced placement. However, the acceptance of these scores for credit is at the discretion of the individual college. Check with the colleges or universities to which you are applying to see what AP® scores they accept for college credit or advanced placement.

Those taking the AP® Calculus BC exam will receive a Calculus AB subscore for the part of the BC exam that covers AB topics. About 60% of the questions on the BC exam specifically test AB concepts.

For all students who took the AB and BC exams worldwide, the results for the years 2015–2019 follow:

AB Exam Score Distributions, 2015–2019

Score	2015	2016	2017	2018	2019
5	21.8%	24.8%	18.7%	19.4%	19.1%
4	17.0%	18.3%	18.0%	17.3%	18.7%
3	18.6%	17.4%	20.8%	21.0%	20.6%
2	10.3%	9.7%	22.0%	22.4%	23.3%
1	32.3%	40.7%	20.4%	20.0%	18.3%
Average	2.86	2.96	2.93	2.94	2.97

BC Exam Score Distributions, 2015–2019

Score	2015	2016	2017	2018	2019
5	45.4%	48.5%	42.6%	40.4%	43.0%
4	16.4%	15.4%	18.1%	18.6%	18.5%
3	18.0%	17.2%	19.9%	20.7%	19.5%
2	5.5%	5.8%	14.1%	14.6%	13.9%
1	14.8%	13.2%	5.3%	5.6%	5.2%
Average	3.72	3.80	3.78	3.74	3.81

What does this all mean? Why do so many more students who take the AP® Calculus BC exam score a 5? It has to do with the level of the student taking the BC exam. It doesn't mean they're smarter, but rather that they've already been through the Calculus AB course, which lays the foundation for the BC-level exam. Thus, it stands to reason that the BC exam would enjoy greater success. After all, almost 85% who took the BC exam scored a 3 or higher on the AB portion of the BC test.

The percentage of total exam questions you need to answer to earn a 5 on either the AB or BC exam may surprise you. Past exams released by the College Board suggest that it is usually about 65%. In the past, earning a 4 has required about 54%, a 3 about 39%, and a 2 just 28%.

Now, let us help you earn a top score. If you study this *Crash Course* book and pay attention during the school year, you will likely be pleasantly surprised how well you do on the exam.

IV. STRATEGIES FOR SCORING HIGH

One of the best ways to prepare for your exam is to research past exams. Although the AP® Calculus exams may change from year to year, it still makes sense to go back to previous tests and answer the questions. The single most important aspect of scoring high on any standardized test is to have complete familiarity with the questions that will be asked. You may not find the exact questions, but you will find those that are similar in content to questions you will encounter on your exam. FRQs are posted on the College Board website (*apcentral.collegeboard.org*). Also on the site is a suite of new resources, including a library of more than 15,000 questions, some never previously released by the College Board. The more questions you answer, the better prepared you will be for the questions come exam day.

On the actual exam, make sure you write clearly. This sounds obvious, but if the AP® Reader scoring your exam cannot read your answer, you will lose credit. We suggest that you cross out your incorrect work rather than erase it.

Along those lines, keep in mind that because you will be graded on your method of calculations, make sure you show all of your work. Clearly identify functions, graphs, tables, or any other items that you've included in order to reach your conclusions.

When you are given a graph, pay attention to its label so you know exactly what function you are being shown. And, if you choose to supply a graph, chart, or table in your answer, be sure it is labeled as well.

You need not simplify numeric or algebraic expressions. An answer of $\frac{1}{\sqrt{2}}$ is adequate rather than the simplified $\frac{\sqrt{2}}{2}$. But if you simplify incorrectly, you will only receive partial credit so don't waste your time simplifying. Also, decimal approximations should be correct to three decimal places with the exception of some series error problems that call for extreme accuracy. In that case, the amount of accuracy necessary will be specified.

V. USING SUPPLEMENTAL INFORMATION

This *Crash Course* contains everything you need to score well on the AP® Calculus AB and BC exams. The AP® Calculus Course and Exam Description published by the College Board can also be a very useful tool in your studies (*www.collegeboard.org*).

Good luck on your AP® Calculus exams!

PART II

FUNCTIONS, GRAPHS, AND LIMITS

Review of Essential Graphs

I. ANALYSIS OF GRAPHS

A. Basic Functions—you need to know how to graph the following functions and any of their transformations by hand.

1. Polynomials, absolute value, square root functions

i. $y = x$ Linear function Domain: $(-\infty, \infty)$ Range: $(-\infty, \infty)$ Odd	ii. $y = x^2$ Quadratic function Concave up Min. point (0, 0) Domain: $(-\infty, \infty)$ Range: $[0, \infty)$ Even	iii. $y = x^3$ Cubic function Increasing Domain: $(-\infty, \infty)$ Range: $(-\infty, \infty)$ No min/max Odd
$y = f(x)$	$y = f(x)$	$y = f(x)$

iv. $y = \lvert x \rvert = \begin{cases} x, & x \geq 0 \\ -x, & x < 0 \end{cases}$ <u>Absolute value</u> function V-shape opens upward Not differentiable at $x = 0$ Domain: $(-\infty, \infty)$ Range: $[0, \infty)$ Even	v. $y = \sqrt{x}$ <u>Square root</u> function Strictly increasing Not differentiable at $x = 0$ Domain: $[0, \infty)$ Range: $[0, \infty)$
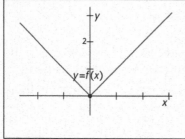	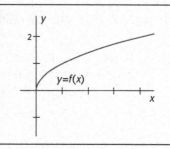

2. Trigonometric functions

i. $y = \sin(x)$ x-intercepts: $x = n\pi$, n is an integer y-intercept: $y = 0$ Odd	ii. $y = \cos(x)$ x-intercepts: $x = (2n+1)\dfrac{\pi}{2}$, n is an integer y-intercept: $y = 1$ Even
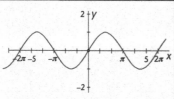	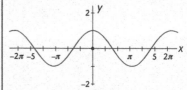

iii. $y = \tan(x)$
x-intercepts: $x = n\pi$, n is an integer

vertical asymptotes: $x = (2n + 1)\dfrac{\pi}{2}$, n is an integer
y-intercept: $y = 0$
Odd

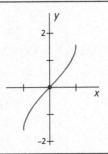

3. Inverse trigonometric functions and their domain and range

i. $y = \sin^{-1}(x)$ Domain: [−1, 1] Range: $\left[-\dfrac{\pi}{2}, \dfrac{\pi}{2}\right]$ Strictly increasing	ii. $y = \cos^{-1}(x)$ Domain: [−1, 1] Range: [0, π] Strictly decreasing
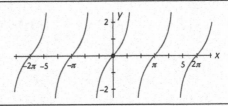	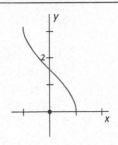

iii. $y = \tan^{-1}(x)$
Domain: (−∞, ∞) Range: $\left(-\dfrac{\pi}{2}, \dfrac{\pi}{2}\right)$

Horizontal Asymptotes: $y = \pm\dfrac{\pi}{2}$, strictly increasing

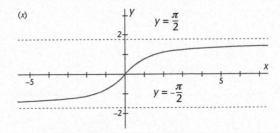

4. Exponential and Natural Logarithmic functions

i. $y = e^x$	ii. $y = \ln(x)$
This is the inverse of $y = \ln(x)$ Strictly increasing x-intercepts: none y-intercept: $y = 1$ horizontal asymptote: $y = 0$	This is the inverse of $y = e^x$ Strictly increasing x-intercept: $x = 1$ y-intercepts: none vertical asymptote: $x = 0$
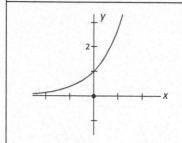	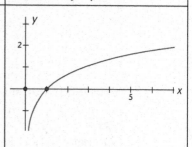

5. Rational functions

i. $y = \dfrac{1}{x}$	ii. $y = \dfrac{1}{x^2}$
Undefined at $x = 0$ Not differentiable at $x = 0$ x-intercepts: none y-intercepts: none horizontal asymptote: $y = 0$ vertical asymptote: $x = 0$ $\lim_{x \to 0} f(x)$ *dne* (does not exist)	Undefined at $x = 0$ Not differentiable at $x = 0$ x-intercepts: none y-intercepts: none horizontal asymptote: $y = 0$ vertical asymptote: $x = 0$ $\lim_{x \to 0} f(x) = \infty$
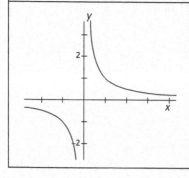	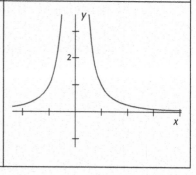

6. Piecewise functions

i. $y = \begin{cases} x, & x \leq 0 \\ x^2, & x > 0 \end{cases}$	ii. $y = \begin{cases} x+1, & x \leq 0 \\ x^2, & x > 0 \end{cases}$
Continuous at $x = 0$ (because the *y* value of each piece at $x = 0$ is the same, $y = 0$) *Not differentiable* at $x = 0$ (because the derivatives of the pieces at $x = 0$ are not equal)	*Discontinuous* at $x = 0$ (because here the *y* value of the left side of the graph is 1 ($0 + 1 = 1$) and the *y* value of the right side of the graph does not exist.) *Not differentiable* at $x = 0$ (since it's not continuous there)
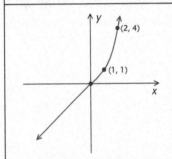	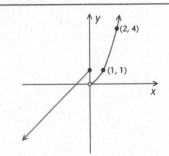

iii. $y = \begin{cases} x^3, & x \leq 0 \\ x^2, & x > 0 \end{cases}$

Continuous at $x = 0$ (because the *y* value of each piece at $x = 0$ is the same, $y = 0$)
Differentiable at $x = 0$ (because the two pieces have equal derivatives at $x = 0$)

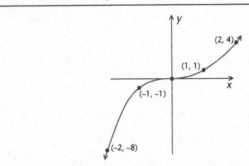

7. Circle Equations

   i. Upper semicircle with radius *a* and center at the origin: $y = \sqrt{a^2 - x^2}$. This is a function. For example, $y = \sqrt{9 - x^2}$

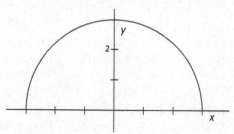

   ii. Lower semicircle with radius *a* and center at the origin: $y = -\sqrt{a^2 - x^2}$. This is a function. For example, $y = -\sqrt{9 - x^2}$

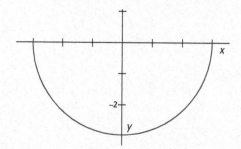

   iii. Circle with radius *a* and center at the origin: $x^2 + y^2 = a^2$. This is not a function since some *x*-values correspond to more than one *y*-value. For example, $x^2 + y^2 = 9$

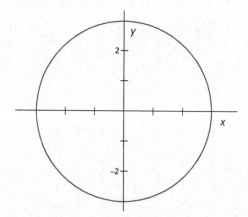

iv. Circle with radius a and center at (b, c): $(x - b)^2 + (y - c)^2 = a^2$. This is not a function either. For example, $(x - 2)^2 + (y + 3)^2 = 9$

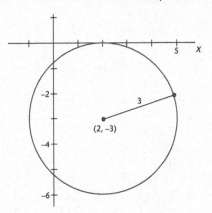

8. Summary of Basic Transformations of Functions

   i. Making changes to the equation of $y = f(x)$ will result in changes in its graph. The following transformations occur most often.

Transformation	$g(x)$ is obtained when	Example
$y = -f(x)$	$y = f(x)$ reflects across the x-axis	$f(x) = x^2$ $\quad g(x) = -x^2$

$y = f(x + a)$, $a > 0$	$y = f(x)$ translates a units left (horizontal shift)	$g(x) = (x + 3)^2$ $f(x) = x^2$
$y = f(x - a)$, $a > 0$	$y = f(x)$ translates a units right (horizontal shift)	$f(x) = x^2$ $g(x) = (x - 2)^2$
$y = f(x) + a$, $a > 0$	$y = f(x)$ translates a units up (vertical shift)	$g(x) = x^2 + 1$ $f(x) = x^2$
$y = f(x) - a$, $a > 0$	$y = f(x)$ translates a units down (vertical shift)	$f(x) = x^2$ $g(x) = x^2 - 4$

$y = af(x)$ $0 \le a < 1$	$y = f(x)$ widens	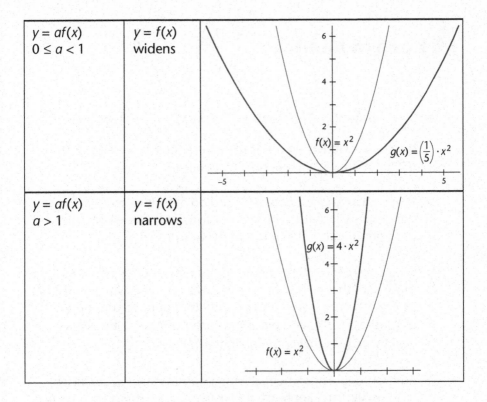
$y = af(x)$ $a > 1$	$y = f(x)$ narrows	

ii. For trigonometric functions, $f(x) = a \sin(bx + c) + d$ or $f(x) = a \cos(bx + c) + d$, a is the amplitude (half the height of the function), b is the frequency (the number of times that a full cycle occurs in a domain interval of 2π units), $\frac{c}{b}$ is the horizontal shift and d is the vertical shift.

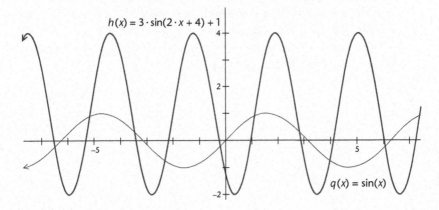

Keep in Mind...

➤ $\dfrac{1}{\sin(x)} \neq \sin^{-1}(x)$. The reciprocal of sin(x), $\dfrac{1}{\sin(x)}$, is equivalent to csc(x), whereas sin⁻¹(x) is the inverse of sin(x), which is the reflection of $y = \sin(x)$ across the line $y = x$.

➤ When changing a function by adding a positive constant to *x*, the graph will shift to the *left*, not the right. The graph shifts to the right *a* units when *a* is subtracted from *x*.

➤ When graphing a function on the calculator (TI-83 or TI-84), make sure that all the plots are turned off; otherwise you risk getting an error and not being able to graph. To turn off the plots, press Ⓨ ⊜ and place the cursor on the plot you want to deactivate (whichever is highlighted). Press (Enter).

➤ An even-degree polynomial with a positive leading coefficient has *y*-values that approach infinity as $x \to \pm\infty$ (both ends go up). If the polynomial has a negative leading coefficient, its *y*-values approach negative infinity as $x \to \pm\infty$ (both ends go down).

➤ An odd-degree polynomial with a positive leading coefficient has *y*-values that approach infinity as $x \to \infty$ and *y*-values that approach negative infinity as $x \to -\infty$ (the right end goes up and the left end goes down). If the polynomial has a negative leading coefficient, its *y*-values approach negative infinity as $x \to \infty$; as $x \to -\infty$ its *y*-values approach positive infinity (the right end goes down and the left end goes up).

CHAPTER 2
PRACTICE PROBLEMS

(See solutions on page 217)

For each of these functions, draw the mother function and the given function on the same set of axes.

1. $y = \dfrac{1}{2}(x+1)^3 - 3$

2. $y = 2\,|\,3x+4\,|$

3. $y = \sqrt{x-6} + 1$

4. $y = -\dfrac{3}{x} + 2$

5. $y = e^{x+2} - 1$

6. $y = \ln(4-x)$

Limits and Continuity

I. MEANING OF LIMIT

A. The limit of a function, $y = f(x)$, as x approaches a number or $\pm\infty$, represents the value that y <u>approaches</u>.

B. The *left-hand* limit, $\lim\limits_{x \to a^-} f(x) = L$, states that as x approaches a, from the <u>*left* of a</u>, $f(x)$ approaches L.
   The *right-hand* limit, $\lim\limits_{x \to a^+} f(x) = L$, states that as x approaches a, from the <u>*right* of a</u>, $f(x)$ approaches L.

C. The expression $\lim\limits_{x \to a} f(x) = L$ states that as x approaches a, <u>simultaneously from the *left and right* of a</u>, $f(x)$ approaches L.

D. The limit of a function at a point exists if and only if the left- and right-hand limits exist *and* are equal.
   Symbolically, if $\lim\limits_{x \to a^-} f(x) = L$ and $\lim\limits_{x \to a^+} f(x) = L$ then $\lim\limits_{x \to a} f(x) = L$.
   The converse is also true.

E. If the left- and right-hand limits are not equal at a given x value, then the limit at the given x value does not exist.
   Symbolically, if $\lim\limits_{x \to a^-} f(x) \neq \lim\limits_{x \to a^+} f(x)$ then $\lim\limits_{x \to a} f(x)$ does not exist.
   The converse is also true.

II. EVALUATING LIMITS ALGEBRAICALLY

A. Generally, $\lim\limits_{x \to a} f(x) = f(a)$. That is, to evaluate the limit of a function algebraically, substitute x with the value x approaches.

(If $x \to \infty$ or $x \to -\infty$, substitute x with values that are very large or very small, respectively.)

1. If $\lim\limits_{x \to a} f(x) = \dfrac{b}{0}$, $b \neq 0$, then take the left- and right-hand limits separately to see if they are the same or not. (In this case, $x = a$ is a *vertical asymptote* of $y = f(x)$.) For instance, $\lim\limits_{x \to 0} \dfrac{1}{x} = \dfrac{1}{0}$ after substituting 0 for x. Since the left-hand limit, $\lim\limits_{x \to 0^-} \dfrac{1}{x} = -\infty$, and the right-hand limit, $\lim\limits_{x \to 0^+} \dfrac{1}{x} = \infty$, are not the same, $\lim\limits_{x \to 0} \dfrac{1}{x}$ does not exist. Similarly, $\lim\limits_{x \to 0} \dfrac{1}{x^2} = \dfrac{1}{0}$. However, the left-hand limit, $\lim\limits_{x \to 0^-} \dfrac{1}{x^2} = \infty$, and the right-hand limit, $\lim\limits_{x \to 0^+} \dfrac{1}{x^2} = \infty$, so $\lim\limits_{x \to 0} \dfrac{1}{x^2} = \infty$.

2. Indeterminate forms:

$\lim\limits_{x \to a} f(x)$ *or* $\lim\limits_{x \to \pm\infty} f(x)$	Method
$\dfrac{0}{0}$	L'Hôspital's Rule (see chapter 5) or simplify by factoring first. For example, using L'Hôspital's Rule, $\lim\limits_{x \to 2} \dfrac{x-2}{x^2-4} = \lim\limits_{x \to 2} \dfrac{1}{2x} = \dfrac{1}{4}$ or, by factoring, $\lim\limits_{x \to 2} \dfrac{x-2}{x^2-4} = \lim\limits_{x \to 2} \dfrac{1}{x+2} = \dfrac{1}{4}$.
$\dfrac{\infty}{\infty}$	L'Hôspital's rule or if $x \to \infty$ or $x \to -\infty$ consider only the highest-power terms when taking the limit since the rest are negligible. For example, by L'Hôspital's Rule, $\lim\limits_{x \to \infty} \dfrac{x^2-3x+2}{x^2+2x-4} = \lim\limits_{x \to \infty} \dfrac{2x}{2x} = 1$ or, by considering only the highest-power terms, $\lim\limits_{x \to \infty} \dfrac{-4x^2-3x+2}{3x^2+2x-4} = \lim\limits_{x \to \infty} \dfrac{-4x^2}{3x^2} = \dfrac{-4}{3}$.

III. EVALUATING LIMITS GRAPHICALLY

Common limit concepts.

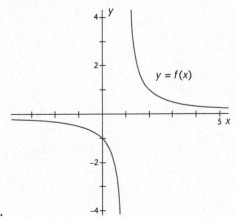

1.

$\lim\limits_{x \to 1} f(x)$ does not exist (*dne*) because the left-hand limit is $-\infty$ and the right-hand limit is ∞.

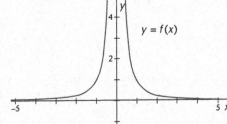

2.

$\lim\limits_{x \to 0} f(x) = \infty$ because the left- and right-hand limits are both ∞.

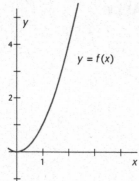

3.

$\lim\limits_{x \to 1} f(x) = 1$ and $f(1) = 1$. The function value and the limit at $x = 1$ are equal.

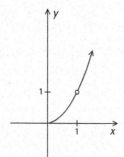

4.

$\lim\limits_{x \to 1} f(x) = 1$ and $f(1)$ *dne*. The function is undefined at $x = 1$ but the limit exists there.

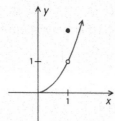

5.

$\lim\limits_{x \to 1} f(x) = 1$ and $f(1) = 2$. The function value is not equal to the limit at $x = 1$.

IV. OTHER KEY LIMITS TO UNDERSTAND

A. $\lim\limits_{x \to 0} \dfrac{\sin(x)}{x} = 1$ and $\lim\limits_{x \to 0} \dfrac{\cos(x) - 1}{x} = 0$ are the most common trigonometric limits.

B. Basic definitions of e: $\lim\limits_{x \to \infty} \left(1 + \dfrac{1}{x}\right)^{x} = e$ or $\lim\limits_{x \to 0}(1 + x)^{\frac{1}{x}} = e$

Keep in Mind...

➤ The answer to a limit question can only be one of the following: a number, $-\infty$, ∞, or "does not exist."

➤ Try to imagine the graph of a function when taking the function's limit.

➤ The limit of a function as x approaches a may or may not be the same as the value of the function at $x = a$.

➤ If a function's limit exists, it must equal a number. A function's limit does not exist in two cases: when the limit from the left and right of the x-value are unequal, or when the y-values approach $\pm\infty$.

➤ Beware the indeterminate forms!

V. ASYMPTOTES

A. Asymptotes are vertical or horizontal lines (the AP® Calculus exams do not include oblique asymptotes) that a graph approaches. Polynomials do not have asymptotes. The functions that most commonly have asymptotes are rational functions,

$y = \dfrac{f(x)}{g(x)}$ as well as other functions such as $y = e^x$, $y = \ln(x)$, $y = \tan(x)$, $y = \cot(x)$, $y = \sec(x)$, $y = \csc(x)$, and their transformations.

1. Vertical asymptotes are vertical lines that a graph only approaches but *never intersects*. (Well, almost never! See a rare example of an exception below.)

   i. $x = k$ is a vertical asymptote of $y = f(x)$ if and only if

   $$\lim_{x \to k^-} f(x) = \pm\infty, \text{ or } \lim_{x \to k^+} f(x) = \pm\infty, \text{ or both.}$$

   ii. For a rational function $y = \dfrac{f(x)}{g(x)}$, the equation of a vertical asymptote is $x = k$ if and only if $g(k) = 0$ and $f(k) \neq 0$; if, however, $f(a) = g(a) = 0$, then, in most cases, there is a removable discontinuity (a hole, not a vertical asymptote) at $x = a$. An example of an exception to the rule is $y = \dfrac{x}{|x|}$ which has an irremovable (nonremovable) discontinuity at $x = 0$ but no vertical asymptote. Also, the function $y = \dfrac{x-1}{x^2 - 1}$ has one hole (at $x = 1$, where both numerator and denominator are 0) and one vertical asymptote (at $x = -1$, where only the denominator is 0). See graphs below:

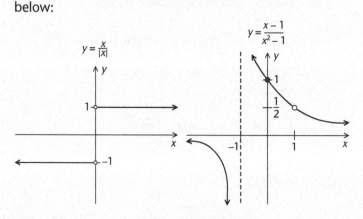

$$y = \frac{x}{|x|}$$

$$y = \frac{x-1}{x^2-1}$$

   iii. A function can have an infinite number of vertical asymptotes (for example, $y = \tan(x)$)

Rare example of a graph intersecting its vertical asymptote:

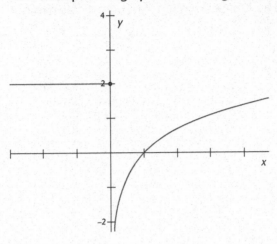

$$f(x) = \begin{cases} 2, & x \le 0 \\ \ln(x), & x > 0 \end{cases}$$

The vertical asymptote for $f(x)$ is $x = 0$ and $f(0) = 2$; thus the point $(0, 2)$ is on the vertical asymptote.

2. Horizontal asymptotes are horizontal lines that a graph approaches and *may intersect*.

   i. $y = k$ is a horizontal asymptote for $y = f(x)$ if and only if $\lim\limits_{x \to \infty} f(x) = k$, or $\lim\limits_{x \to -\infty} f(x) = k$, or both. Horizontal asymptotes give us an idea of the function's end behavior (as $x \to \pm\infty$). The graph of $y = \dfrac{x}{e^{x^2}}$, below, has only one horizontal asymptote, $y = 0$, since $\lim\limits_{x \to \infty} \dfrac{x}{e^{x^2}} = 0$ and $\lim\limits_{x \to -\infty} \dfrac{x}{e^{x^2}} = 0$. Note that the graph intersects its asymptote at $(0, 0)$.

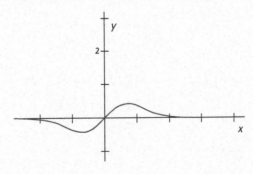

ii. A function can have at most two horizontal asymptotes. The function $y = \tan^{-1}(x)$, below, has two horizontal asymptotes, $y = \pm\dfrac{\pi}{2}$ since $\lim\limits_{x \to \infty} \tan^{-1}(x) = \dfrac{\pi}{2}$ and $\lim\limits_{x \to -\infty} \tan^{-1}(x) = -\dfrac{\pi}{2}$.

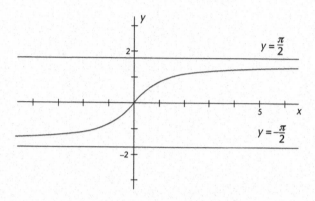

iii. A function which only has one horizontal asymptote does not have to approach this asymptote on both ends! The function $y = \dfrac{x}{e^x}$, below, has only one horizontal asymptote, $y = 0$ since $\lim\limits_{x \to \infty} \dfrac{x}{e^x} = 0$ and $\lim\limits_{x \to -\infty} \dfrac{x}{e^x} = -\infty$ and so it only approaches this asymptote as $x \to \infty$.

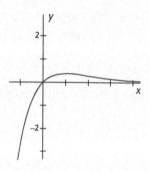

VI. UNBOUNDED BEHAVIOR

A. If a function, $y = f(x)$, approaches positive infinity either as $x \to a$ or as $x \to \pm\infty$, the function is said to increase without bound. Similarly, if a function, $y = f(x)$, approaches negative infinity either as $x \to a$ or as $x \to \pm\infty$, the function is said to decrease without bound.

Keep in Mind...

➤ A graph *might* cross its horizontal asymptote. A graph is very unlikely to cross its vertical asymptote.

➤ Do not confuse vertical/horizontal asymptotes with vertical/horizontal tangent lines.

➤ When finding a vertical asymptote, find the root of the denominator and then make sure that it is not also a root of the numerator. To be safe, simplify a rational function completely before finding its vertical asymptotes.

➤ Not only rational functions have asymptotes. Functions such as $y = \ln(x)$, $y = e^x$, $y = \tan(x)$ and their transformations have asymptotes as well.

VII. CONTINUITY OF FUNCTIONS

A. A function is either continuous (no breaks whatsoever) or discontinuous at certain points.
 1. A function $y = f(x)$ is continuous at a point, $x = a$, if and only if $\lim\limits_{x \to a^-} f(x) = \lim\limits_{x \to a^+} f(x) = f(a)$. Simply put, this states that for $y = f(x)$ to be continuous at $x = a$, the limits of the function

from the left and right of *a* must be equal to each other and also equal to the value of the function at $x = a$.

2. All polynomials are continuous.

3. Some of the most common discontinuous functions come in the form of rational functions, piecewise functions and $y = \tan(x)$, $y = \cot(x)$, $y = \sec(x)$, $y = \csc(x)$ and their transformations.

4. A *removable discontinuity* occurs when an otherwise continuous graph has a point (or more) missing. That is, $y = f(x)$ has a removable discontinuity at $x = a$ if and only if $\lim\limits_{x \to a^-} f(x) = \lim\limits_{x \to a^+} f(x) = L$ but $f(a) \neq L$ or $f(a)$ does not exist.

   i. The function $f(x) = \begin{cases} x, & x \neq 1 \\ 2, & x = 1 \end{cases}$ has a removable discontinuity

   at $x = 1$ since $\lim\limits_{x \to 1^-} f(x) = \lim\limits_{x \to 1^+} f(x) = 1$ and $f(1) \neq 1$. In this case, $f(1) = 2$.

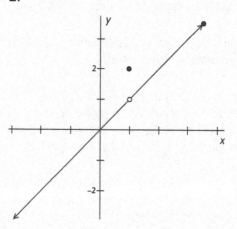

   ii. The function $y = \dfrac{\sin(x)}{x}$ has a removable discontinuity at

   $x = 0$ since $\lim\limits_{x \to 0^-} \dfrac{\sin(x)}{x} = \lim\limits_{x \to 0^+} \dfrac{\sin(x)}{x} = 1$ but $f(0) \neq 1$. In this case, $f(0)$ does not exist.

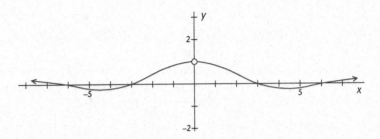

iii. A removable discontinuity is one that can be "filled in" (or removed) if the function is appropriately redefined. To remove the discontinuity of the function in part *i*, the function can be redefined as $f(x) = x$. To remove the discontinuity in part *ii*, the function can be redefined as

$$y = \begin{cases} \dfrac{\sin(x)}{x}, & x \neq 0 \\ 1, & x = 0 \end{cases}$$

5. A *nonremovable discontinuity* occurs at step breaks in the graph or at vertical asymptotes. That is, $y = f(x)$ has a nonremovable discontinuity at $x = a$ if and only if $\lim\limits_{x \to a^-} f(x) \neq \lim\limits_{x \to a^+} f(x)$ or if one or both of these limits is $\pm\infty$.

i. The function $y = \dfrac{1}{x}$ has a nonremovable discontinuity at $x = 0$ (also a vertical asymptote there) because the function cannot be redefined so that it will be continuous there. This is also called an infinite discontinuity.

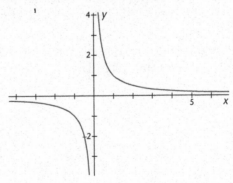

ii. The function $y = \dfrac{x}{|x|}$ has a nonremovable discontinuity at

$x = 0$ (a step break, not a vertical asymptote) because we cannot redefine it so that it will become continuous there. This is also called a jump discontinuity.

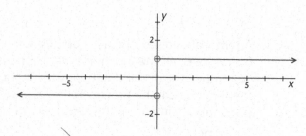

6. If a function is continuous, it does not have to be differentiable! (A function is continuous at a cusp or corner, yet it is not differentiable there; see more in Chapter 4, but here, differentiable curves are smooth.)

Keep in Mind...

➤ Loosely speaking, a continuous function is one which can be drawn without lifting the pencil off the paper.

➤ Continuity does not imply differentiability. Differentiability does imply continuity.

CHAPTER 3
PRACTICE PROBLEMS

(See solutions on page 219)

1. $\lim\limits_{x \to \infty} \dfrac{1 - 3x + 6x^2 - x^{10}}{2 + 4x^4 - 8x^7 + 8x^{10}} =$

 (A) 0

 (B) ∞

 (C) $-\infty$

 (D) $-\dfrac{1}{8}$

2. $\lim\limits_{x \to \infty} \dfrac{\sin x}{x} =$

 (A) 0

 (B) ∞

 (C) 1

 (D) Does not exist

3. $\lim\limits_{x \to 0^+} \ln(x) =$

 (A) 1

 (B) -1

 (C) 0

 (D) $-\infty$

4. $\lim\limits_{x \to 3} \dfrac{\sqrt{x+6} - 3}{x - 3} =$

 (A) 1

 (B) -1

 (C) $\dfrac{1}{6}$

 (D) ∞

5. $\lim\limits_{x \to 1} \dfrac{3}{x - 1} =$

 (A) 1

 (B) ∞

 (C) $-\infty$

 (D) Does not exist

37

Find the horizontal and vertical asymptotes of the following functions:

6. $f(x) = \dfrac{3x^2 - 9x}{x^2 - 9}$ $\dfrac{3x\,(x-3)}{(x+3)(x-3)}$ $x \neq 3$

7. $f(x) = \dfrac{x^3 + 3x^2 - 1}{4 - x^3}$

8. What conclusion can you draw about the asymptotes of the nonlinear function $f(x)$ if:

(A) $\displaystyle\lim_{x \to \infty} f(x) = 7$ horizontal

(B) $\displaystyle\lim_{x \to -\infty} f(x) = -\infty$

(C) $\displaystyle\lim_{x \to 4} f(x) = \infty$ vertical

9. Find the x-values for which $f(x) = \dfrac{2}{\sqrt{1-x}}$ is continuous.

10. Find the discontinuities of $f(x) = \dfrac{x^2 + 5x + 6}{x^2 - 4}$ and categorize them as removable or nonremovable.

11. Find all x-values for which $f(x) = \begin{cases} 2 - x, & x < -1 \\ \dfrac{1}{x}, & -1 \le x \le 2 \\ \dfrac{1}{2}, & x > 2 \end{cases}$ is discontinuous.

PART III
DIFFERENTIATION

Differentiation: Definition, Principles and Techniques

I. DERIVATIVES

A. Meaning of Derivative

The <u>derivative</u> of a function is a formula for its <u>slope</u>. A linear function has a constant derivative since its slope is the same at every point. The derivative of a function at a point is the slope of its tangent line at that point. Non-linear functions have derivatives that depend on the value of x since the slope is constantly changing.

1. Local linearity or linearization—when asked to find the linearization of a function at a given x-value or when asked to find an approximation to the value of a function at a given x-value using the tangent line, this means finding the equation of the tangent line at a "nice" x-value in the vicinity of the given x-value, substituting the given x-value into it and solving for y.

   i. For example, approximate $\sqrt{4.02}$ using the equation of a tangent line to $f(x) = \sqrt{x}$. We'll find the equation of the tangent line to $f(x) = \sqrt{x}$ at $x = 4$ (this is the 'nice' x-value mentioned earlier). What makes it nice is that it is close to 4.02 and that $\sqrt{4} = 2$). Since $f'(x) = \dfrac{1}{2\sqrt{x}}$, $f'(4) = \dfrac{1}{2\sqrt{4}} = \dfrac{1}{4}$ so, $m = \dfrac{1}{4}$. Also, $f(4) = 2$. Substituting these values into the equation of the tangent line,

   $y = mx + b \rightarrow 2 = \dfrac{1}{4}(4) + b \rightarrow b = 1$ so the equation

   of the tangent line is $y = \dfrac{1}{4}x + 1$. Substituting $x = 4.02$,

$y = 2.005$. A more accurate answer (using the calculator) is $\sqrt{4.02} = 2.004993766$. The linear approximation, 2.005, is very close to this answer. This works so well because the graph and its tangent line are very close at the point of tangency, thus making their y-values very close as well. If you use the tangent line to a function at $x = 4$ to approximate the function's value at $x = 9$, you will get a very poor estimate because at $x = 9$, the tangent line's y-values are no longer close to the function's y-values.

ii. The slope of the secant on (a, b), is often used to approximate the value of the slope at a point inside (a, b). For instance, given the table of values of $f(x)$ below, and given that $f(x)$ is continuous and differentiable, approximate $f'(3)$. You will not be told to use the slope of the secant between two points containing $x = 3$, you'll just have to know to do this. So, $m_{x=3} \approx \dfrac{2.9 - 1.3}{5 - 2} = .5\overline{3}$ or $m_{x=3} \approx \dfrac{1.6 - 1.3}{3 - 2} = 1.3$. There can be different answers since this is only an approximation.

x	$f(x)$
2	1.3
3	1.6
5	2.9
6	2.8

B. Notation of Derivative and Common Terms Used to Describe It

1. Common notations: $f'(x)$, y', $\dfrac{dy}{dx}$, $\dfrac{d}{dx}(f(x))$

2. Common terms to describe the derivative: instantaneous rate of change, change in y with respect to x, slope.

C. Definition of Derivative

1. Derivative as a function: $f'(x) = \lim\limits_{h \to 0} \dfrac{f(x+h) - f(x)}{h}$

2. Derivative at a point, $x = a$: $f'(a) = \lim\limits_{x \to a} \dfrac{f(x) - f(a)}{x - a}$ (Notice that this is equivalent to $(m_{tangent}$ at $x = a) = \lim\limits_{x \to a} (m_{secant}$ between x and $a)$.

This is to say that the slope of the tangent line at $x = a$ is equal to the limit of the slope of the secant line between $x = a$ and any other x-value, as the x-value approaches a.

D. Existence of Derivative at a Point

A function's derivative does not exist at points where the function has a discontinuity, corner, cusp, vertical asymptote or vertical tangent.

1. $y = f(x)$

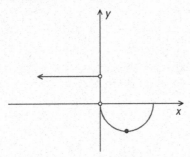

$f'(0)$ does not exist because $f(x)$ is <u>discontinuous</u> at $x = 0$.

2. $y = g(x)$

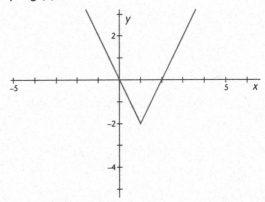

$g'(1)$ does not exist because $g(x)$ has a <u>corner</u> at $x = 1$ because the left and right derivatives are not equal.

3. $y = h(x)$

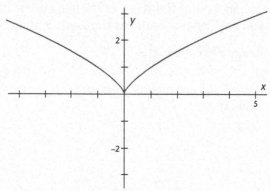

$h'(0)$ does not exist because at $x = 0$ $h(x)$ has a <u>cusp</u> because $\lim\limits_{x \to 0^-} h(x) = -\infty$ and $\lim\limits_{x \to 0^+} h(x) = \infty$.

4. $y = s(x)$

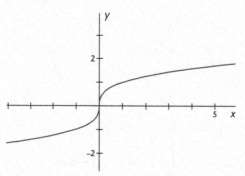

$s'(0)$ does not exist because $s(x)$ has a <u>vertical tangent</u> at $x = 0$

E. Properties of $f(x) = e^x$ and $g(x) = \ln(x)$
1. $e^{\ln(x)} = x$
2. $\ln(e^x) = x$
3. $\ln(xy) = \ln(x) + \ln(y)$
4. $\ln\left(\dfrac{x}{y}\right) = \ln(x) - \ln(y)$
5. $\ln(x^y) = y \ln(x)$
6. $\ln(1) = 0$
7. $\ln(0)$ does not exist

DERIVATIVE RULES

A. Algebraic Expressions

y	y'	Rule Name	Examples
k	0	Constant Rule	$y = 3 \rightarrow y' = 0$
ax^n	anx^{n-1}	Power Rule	$y = 3x^5 - 2x^4$ $\rightarrow y' = 15x^4 - 8x^3$
$f(x) \cdot g(x)$	$f(x)\,g'(x) + g(x)f'(x)$	Product Rule	$y = x^2(x^3 - 3x)$ $\rightarrow y' = x^2(3x^2 - 3)$ $+ (x^3 - 3x)(2x)$
$\dfrac{f(x)}{g(x)}$	$\dfrac{g(x)f'(x) - f(x)g'(x)}{[g(x)]^2}$	Quotient Rule	$y = \dfrac{x^2}{2x+1}$ $\rightarrow y' = \dfrac{(2x+1)(2x) - x^2(2)}{(2x+1)^2}$ $= \dfrac{2x^2 + 2x}{(2x+1)^2}$
$f(g(x))$	$f'(g(x))g'(x)$	Chain Rule	$y = (x^2 + 3x + 7)^5$ $\rightarrow y' = 5(x^2 + 3x + 7)^4(2x+3)$ Here, the "inside function," $g(x) = (x^2 + 3x + 7)$ The "outside function," $f(x) = x^5$.

When taking the derivative of a function, you might have to use more than one of the above rules.

B. The derivative of a cofunction (any trig function starting with the letter C) is negative.

y	$\sin(x)$	$\cos(x)$	$\tan(x)$	$\cot(x)$	$\sec(x)$	$\csc(x)$
y'	$\cos(x)$	$-\sin(x)$	$\sec^2(x)$	$-\csc^2(x)$	$\sec(x)\tan(x)$	$-\csc(x)\cot(x)$

1. The derivatives of the cofunctions are negative.
2. In taking the derivative of most trigonometric functions, you will need to use the chain rule since most will be compositions—sometimes of more than two functions. Here is an example of the derivative of a function of the form $y = f(g(h))$: $y = \sin(\tan(x^2)) \rightarrow y' = \cos(\tan(x^2))\sec^2(x^2)(2x)$.

C. Derivatives of inverse trigonometric functions

y	$\sin^{-1}(x)$	$\cos^{-1}(x)$	$\tan^{-1}(x)$	$\cot^{-1}(x)$	$\sec^{-1}(x)$	$\csc^{-1}(x)$
y'	$\dfrac{1}{\sqrt{1-x^2}}$	$-\dfrac{1}{\sqrt{1-x^2}}$	$\dfrac{1}{1+x^2}$	$-\dfrac{1}{1+x^2}$	$\dfrac{1}{x\sqrt{x^2-1}}$	$-\dfrac{1}{x\sqrt{x^2-1}}$

1. Note that the derivatives of the cofunctions are the negatives of the derivatives of the functions.
2. In most cases, the chain rule is used. For example,
$$y = \sec^{-1}(\cos(3x)) \rightarrow y' = \frac{-3\sin(3x)}{(\cos(3x))\sqrt{\cos^2(3x)-1}}.$$
3. Although the derivative of the inverse csc, sec, and cot are rarely used, students should memorize the derivatives of the $\sin^{-1}x$, $\cos^{-1}x$, and $\tan^{-1}x$ functions.

III. IMPLICIT DIFFERENTIATION

Implicit differentiation means finding y' when the equation given is not explicitly defined in terms of y (that is, it is not of the form $y = f(x)$). In this case you must remember to always use the chain rule when taking the derivative of an expression involving y.

Example 1: Find y' if $x^2 + y^2 = 3$.

Taking derivatives on both sides, $2x + 2yy' = 0 \rightarrow y' = -\dfrac{x}{y}$.

Example 2: Find y' if $x^2y^2 - 3 \ln y = x + 7$.

Taking derivatives on both sides, $x^2(2yy') + y^2(2x) - \dfrac{3}{y}y' = 1$

$\rightarrow 2x^2yy' - \dfrac{3}{y}y' = 1 - 2xy^2$

$\xrightarrow{\text{factor out } y'} y' = \dfrac{1 - 2xy^2}{\left(2x^2y - \dfrac{3}{y}\right)}$

$\xrightarrow{\text{multiply numerator and denominator by } y} y' = \dfrac{y - 2xy^3}{2x^2y^2 - 3}$. Note that the

product rule must be used here when taking the derivative of x^2y^2.

Example 3: Find y' if $x^2 - xy = x + y$.

Taking derivatives on both sides, $2x - (xy' + y) = 1 + y' \rightarrow$
$2x - xy' - y = 1 + y' \rightarrow 2x - y - 1 = y' + xy'$
$\rightarrow y' + xy' = 2x - y - 1 \rightarrow y'(1 + x) = 2x - y - 1 \rightarrow$
$y' = \dfrac{2x - y - 1}{1 + x}$. Note that you must use the product rule

when taking the derivative of xy, and must distribute the negative sign!

IV. DERIVATIVE OF THE INVERSE OF A FUNCTION

A. $y = f(x) \rightarrow \dfrac{d}{dx}(y^{-1}) = \dfrac{1}{f'(f^{-1}(x))}$

  1. An example using the formula: $f(x) = \sqrt{x}$. Since $f' = \dfrac{1}{2\sqrt{x}}$

  and $f^{-1}(x) = x^2$ (for $x > 0$), then, $\dfrac{d}{dx}(y^{-1}) = \dfrac{1}{f'(f^{-1}(x))} =$

  $\dfrac{1}{\dfrac{1}{2\sqrt{x^2}}} = \dfrac{1}{\dfrac{1}{2x}} = 2x$. This is only an illustration of this formula.

Certainly you can find the derivative of the inverse more directly by finding the inverse first and then taking its derivative. In many cases, this is difficult or impossible or simply time-consuming. Generally, to find the derivative of the inverse of a function, switch x and y and find y' implicitly. If asked to evaluate the derivative of the inverse of a function at a point, make sure you know which point you are given, one on the function or one on the inverse. Remember that if (a, b) is a point on a function, then (b, a) is a point on its inverse. The converse is also true.

2. An example without using the formula: given $f(x) = x^3 + 2$, evaluate $(f^{-1}(3))'$. Notice that $x = 3$ is an x-value of the inverse. So, rewrite the function, $y = x^3 + 2$, switch x and y, $x = y^3 + 2$. To find y', take the derivative implicitly:

$1 = 3y^2 y' \rightarrow y' = \dfrac{1}{3y^2}$. The y in this equation is the y of the

inverse. So, since $x = 3$, $y = 1$ (substitute $x = 3$ into $x = y^3 + 2$

to find the y-value) and our final answer is $(f^{-1}(3))' = \dfrac{1}{3}$.

V. DERIVATIVES OF TRANSCENDENTAL FUNCTIONS

A. Natural log

$y = \ln(x) \rightarrow y' = \dfrac{1}{x}, x > 0$. In general, using the chain rule,

$y = \ln(f(x)) \rightarrow y' = \dfrac{f'(x)}{f(x)}, f(x) > 0.$

B. Exponentials

$y = a^x \rightarrow y' = a^x \ln a$. In general, using the chain rule,
$y = a^{f(x)} \rightarrow y' = a^{f(x)} f'(x) \ln a$. A special case of this is $y = e^x$. It is the only function which is equal to its derivative, $(e^x)' = e^x$ other than the trivial $y = 0$.

VI. DERIVATIVES OF PIECEWISE FUNCTIONS

A. For a piecewise function to be differentiable at a break point, it must be continuous at that point, and the derivatives of the pieces at that point must be equal. Remember that differentiability implies continuity, but continuity does not imply differentiability. This is to say that if a function is differentiable at a point, then it is continuous at that point. However, if a function is continuous at a point, it is not necessarily differentiable at that point.

1. $y = \begin{cases} x^2 + 1, & x < 1 \\ 2x, & x > 1 \end{cases}$. This function is not differentiable at $x = 1$

   because it is not continuous there as y is not defined at $x = 1$.

   So, $y' = \begin{cases} 2x, & x < 1 \\ 2, & x > 1 \end{cases}$.

2. $y = \begin{cases} x^2 + 1, & x \leq 1 \\ 2x, & x > 1 \end{cases}$. This function is continuous at $x = 1$ because

   $(1)^2 + 1 = 2(1)$ and the derivatives of the pieces are equal at $x = 1$ ($2x = 2$ at $x = 1$). Therefore, this function is differentiable

   at $x = 1$ and $y' = \begin{cases} 2x, & x \leq 1 \\ 2, & x > 1 \end{cases}$.

3. $y = \begin{cases} x^2, & x \leq 1 \\ 2x, & x > 1 \end{cases}$. This function is not differentiable at $x = 1$

   because, even though the derivatives of the pieces at $x = 1$ are equal, this function is not continuous at $x = 1$. So it is

   not differentiable there. $y' = \begin{cases} 2x, & x < 1 \\ 2, & x > 1 \end{cases}$. Notice that $x = 1$

   was excluded from the domain of the derivative since the derivative does not exist there.

4. $y = \begin{cases} x^2 + 1, & x \leq 1 \\ 2, & x > 1 \end{cases}$. This function is continuous at $x = 1$ but

   not differentiable there since $2x \neq 0$ when $x = 1$. That is, the derivatives of the pieces aren't equal there.

Keep in Mind...

➤ When using the quotient rule, do not switch the order of the terms in the numerator since subtraction is not commutative.

➤ When differentiating a function of the form $y = \dfrac{f(x)}{k}$ where k is a constant, do not use the quotient rule. The derivative is simply $y' = \dfrac{f'(x)}{k}$ since $\dfrac{1}{k}$ is a constant that can be factored out.

➤ Do not confuse $\ln(1) = 0$ with $\ln(0) = 1$. The former is true since $x = 1$ is the x-intercept of $y = \ln(x)$. The latter is false since $x = 0$ is not in the domain of $y = \ln(x)$.

➤ Don't forget to use the product rule in implicit differentiation problems in which you must take the derivative of a product involving both x and y. Also, if such an expression is being subtracted, make sure to distribute the negative sign.

CHAPTER 4
PRACTICE PROBLEMS
(See solutions on page 222)

1. Find $\dfrac{dy}{dx}$ if $y = \dfrac{3x + 4}{1 - 5x}$

2. Find $\dfrac{d}{dx}(\ln(e^{\sqrt{5x+3}}))$

3. Evaluate y' at $x = -1$ if $3x - x^2y = 5y$

4. Find the derivative of the inverse of $y = x^2 - 4x$ at $x = 2$.

5. Use the chart to find $h'(4)$ if $h(x) = f(x)g(x)$.

$f(4)$	$f'(4)$	$g(4)$	$g'(4)$
–8	3	3π	4

6. If $f(x) = \left(\dfrac{2x-1}{2x+1}\right)^5$, find $f'(x)$.

7. Given that $f(2) = -3$, $f'(2) = 6$, $g(2) = 3$, $g'(2) = -2$, $f'(3) = 4$, $g'(-3) = -1$, find $h'(2)$ if $h(x) = f(g(x))$.

8. If $y = \sqrt{\cos x^2}$, find y'.

9. If $x^2 - 3 \ln y + y^2 = 25$, find $\dfrac{dy}{dx}$.

10. If $y = (e^x - 2x - 1)^3$, find $\dfrac{dy}{dx}$.

Contextual Applications of Differentiation

I. MOTION

A. Rectilinear Motion—motion in a straight line

  1. Displacement and distance

    i. Displacement is the distance between a moving object's endpoint and starting point. For instance, if an object moves from the origin two units to the right and back to the origin, its displacement is zero. If it moves from the origin two units to the right and then three units to the left, its total displacement is $2 - 3 = -1$ unit. Symbolically, if $S(t)$ represents the path of an object, then the displacement from $t = a$ to $t = b$ is given by $S(b) - S(a)$.

    ii. Distance is the length of the path traveled by an object. If an object moves from the origin two units to the right and back to the origin, the distance it traveled is 4 units. If it moves from the origin two units to the right and then three units to the left, the distance traveled is 5 units.

  2. Speed, velocity, and acceleration

    i. Speed measures an object's change in *distance* traveled per unit of time.

    ii. Velocity measures an object's *displacement* per unit of time. Symbolically, speed = |velocity|. If $s(t)$ represents the displacement of an object, then $s'(t)$ represents the object's velocity.

    iii. Acceleration measures an object's change in velocity per unit of time. Symbolically, $a(t) = v'(t) = s''(t)$. If the acceleration and velocity of an object have the same signs, then the object is speeding up. If the acceleration and velocity have opposite signs, then the object is slowing down. Think of acceleration and velocity as two forces acting on the object—if they act in the same direction, they increase the object's speed; if they act in different directions, they slow the object down.

Example: The graph of $s(t)$, below, represents an object's displacement. From this graph we can deduce that the object is speeding up in two different ways. Notice that the slope of the tangent line to $s(t)$, that is, the velocity, increases as time increases so the object is speeding up. Or, we can analyze the signs of the velocity and acceleration. In this case, $v(t) > 0$ (since $s(t)$ is increasing) and $a(t) > 0$ (since $s(t)$ is concave up); hence, the object is speeding up.

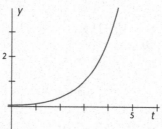

II. RELATED RATES

A. Related rates problems are word problems where you are asked how fast a quantity is changing over time. These problems require you to take the derivative of some quantity with respect to time t, involving implicit differentiation. If the question states that a snowball is melting at 4 in.³/sec., this means that $\frac{dV}{dt} = -4$ in.³/sec. (Look at the units to figure out what variable is being discussed; note that the variable is negative since the snowball's volume is decreasing.)

Test Tip

Common formulas you must memorize for such problems are: the Pythagorean Theorem, proportions in right triangles, area/perimeter of basic geometric figures, and volume of a sphere, cylinder, and cone.

The steps required to solve a related rates problem are:

(a) Create a legend which includes the given information and the variable that you are looking for—it is generally a rate ("how fast is the radius changing with respect to time" means that you are looking for the value of $\frac{dr}{dt}$).

(b) Write the equation that relates the variables given. (If the problem involves the radius, height, and volume of a cylinder, the equation you would use would be that for the volume of a cylinder.)

(c) Take the derivative of the function in part b implicitly with respect to time. Then substitute in the given information and solve for the missing variable.

(d) Double-check that you found the answer to the question being asked. Make sure to include correct units (units squared for area, units cubed for volume). If the related rate problem shows up in the free-response question, write a complete sentence as your answer.

Example: Coffee is poured into a conical cup at a constant rate of 1 in.^{3}/sec. Given that the cup's top radius measures 3 in. and its height is 9 in., find how fast the water level of the coffee in the cup changes when the radius is 2 in.

(a) Create legend: $\frac{dV_{coffee}}{dt} = 1$ in.^{3}/sec. (this is positive because the coffee volume is increasing), $r_{cup} = 3$ in., $h_{cup} = 9$ in.,

$\frac{dh_{coffee}}{dt} = ?$ (Note that there are two cones in this problem: one is the cup and the other is the shape of the coffee in the cup. The coffee in the cup changes dimensions but the cup's dimensions remain constant, so it's important to differentiate between the dimensions of the cup and those of the coffee.)

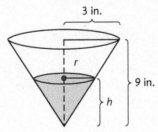

3 in.

9 in.

r

h

(b) $V_{coffee} = \frac{1}{3}\pi r^2_{coffee} h_{coffee}$. The volume in this case is a function of two variables. To make it easier to calculate, we must rewrite it so that it is a function of only one variable, h_{coffee}, since we are looking for $\frac{dh_{coffee}}{dt}$. Using the fact that in similar triangles the ratio of corresponding sides forms a proportion, $\frac{r_{coffee}}{h_{coffee}} = \frac{3}{9} \rightarrow r_{coffee} = \frac{1}{3}h_{coffee} \rightarrow$

$$V_{coffee} = \frac{1}{3}\pi \left(\frac{1}{3}h_{coffee}\right)^2 h_{coffee} \rightarrow V_{coffee} = \frac{1}{27}\pi h^3_{coffee}$$

(c) Take the derivative of the function in part (b) and substitute the given information in order to solve for the unknown quantity.

$$\frac{dV_{coffee}}{dt} = \frac{1}{9}\pi h^2_{coffee}\frac{dh_{coffee}}{dt} \rightarrow 1 = \frac{1}{9}\pi(6)^2\frac{dh_{coffee}}{dt} \rightarrow$$

$\frac{dh_{coffee}}{dt} = \frac{1}{4\pi}$ in./sec. (since $r_{coffee} = \frac{1}{3}h_{coffee}$, when

$r_{coffee} = 2$ in., $h_{coffee} = 6$ in.)

(d) When the radius of the coffee in the cup is 2 in., the coffee level increases at a rate of $\frac{1}{4\pi}$ in./sec.

Keep in Mind...

➤ Don't forget appropriate units!

➤ If a quantity is increasing over time, its derivative with respect to time is positive. If a quantity is decreasing over time, its derivative with respect to time is negative. If a quantity does not change over time, its derivative with respect to time is zero. The converse of each statement is true as well.

➤ Make sure you know the formulas for the volume of a cone, sphere, cube, and cylinder.

III. LOCAL LINEARITY

A. The figure to the right is not a circle but a 24-sided regular polygon (called an icosikaitera). When the number of sides become this large, the length of the sides becomes smaller. The calculus definition of a circle is simply a polygon whose number of sides approaches infinity and whose lengths of sides approach zero. This phenomenon regularly occurs in daily life. Examine guardrails on the side of a curving road and you will see that they do not match the curve of the road but in reality are a series of straight-line rails. Since the straight lines are numerous, they appear to curve. Wooden roller-coaster tracks do this as well. It is quite expensive to create curved rails. So straight rails are used– and when there are many of them, each of them is fairly short and they can simulate a curve.

1. Suppose we have a function $f(x)$ and its tangent line drawn at the point $(a, f(a))$ as shown in the figure to the right. The equation of the tangent line, which we call $L(x)$, is given by the equation

   $$L(x) - f(a) = f'(a)(x - a) \quad \text{or} \quad L(x) = f(a) + f'(a)(x - a).$$

   Near $x = a$, the function and tangent line have nearly the same graph even though one is curved and the other, a straight line. This is called local linearity. If k were a value very near to a, there would be very little difference in the values of $L(k)$ and $f(k)$. In this case, we call the tangent line the linear approximation to the function at $x = a$.

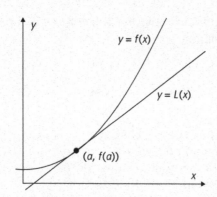

2. *Example:* Use the linear approximation for $f(x) = \sqrt[3]{x}$ at $x = 8$ to approximate $f(x) = \sqrt[3]{8.1}$.

$$f'(x) = \frac{1}{3x^{2/3}} \quad f'(8) = \frac{1}{12} \quad\quad f(8) = 2$$

Tangent line: $y - 2 = \frac{1}{12}(x - 8) \Rightarrow y = \frac{x}{12} + \frac{4}{3} = y = \frac{x+16}{12}$

At $x = 8.1$, $y = \frac{24.1}{12} = 2.0083$

Since 8.1 is close to 8, this approximation should be very close.

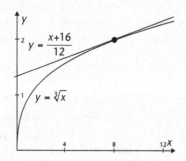

IV. INDETERMINATE FORMS AND L'HÔSPITAL'S RULE

A. In the study of limits, there are problems such as the following:

$$\lim_{x \to 2} \frac{x-2}{x^2-4} \quad \text{and} \quad \lim_{x \to \infty} \frac{1-4x-5x^2}{3x^2-x-4}$$

In the first limit, if we plugged in $x = 2$, we would get 0/0 and in the second limit, "plugging in" infinity gives $-\infty/\infty$. Both of these are called indeterminate forms. In both of these cases, there are rules that are followed that seem to give different results. In the first problem, where we get $\frac{0}{0}$, you have always been told that a fraction with a numerator of zero is zero. However, you have also been told a fraction with a denominator of zero does not exist, so is the limit zero, does it not exist, or is it some other number?

In the second problem where we get $\frac{-\infty}{\infty}$, you have been told that a fraction with a numerator that goes to infinity is infinity. However, you have also been told that a fraction with a denominator that goes to infinity is zero. So is the limit infinity or zero? Or maybe the infinities cancel out and the answer is –1. Add to the conundrum that infinity is not a number and we can't treat it like a number.

1. Some of these problems can be handled with factoring such as:

$$\lim_{x \to 2} \frac{x-2}{x^2-4} = \lim_{x \to 2} \frac{x-2}{(x+2)(x-2)} = \lim_{x \to 2} \frac{1}{x+2} = \frac{1}{4}$$

2. However, this technique will not work with
   $\lim_{x \to 0} \frac{\cos x - 1}{x^2 - x}$ and $\lim_{x \to \infty} \frac{e^x + x}{x^3}$. The first is in the form $\frac{0}{0}$, but no factoring is possible. The second is in the form $\frac{\infty}{\infty}$ and no factoring is possible.

B. L'Hôspital's Rule allows us to evaluate limits involving indeterminate forms.

Suppose we have one of the two following cases:

$$\lim_{x \to a} \frac{f(x)}{g(x)} = \frac{0}{0} \qquad \text{or} \qquad \lim_{x \to a} \frac{f(x)}{g(x)} = \frac{\pm\infty}{\pm\infty}$$

where a can be any real number, infinity, or negative infinity:

Then $\lim_{x \to a} \frac{f(x)}{g(x)} = \lim_{x \to a} \frac{f'(x)}{g'(x)}$.

Example: $\lim\limits_{x \to 2} \left(\dfrac{3x^2 - 7x + 2}{x - 2} \right)$

$$\lim\limits_{x \to 2} \left(\frac{3x^2 - 7x + 2}{x - 2} \right) = \frac{0}{0} \Rightarrow \lim\limits_{x \to 2} \left(\frac{6x - 7}{1} \right) = 5$$

Example: $\lim\limits_{x \to -2} \left(\dfrac{x^3 + x^2 - 8x - 12}{x^3 + 8x^2 + 20x + 16} \right)$

$$\lim\limits_{x \to -2} \left(\frac{x^3 + x^2 - 8x - 12}{x^3 + 8x^2 + 20x + 16} \right) = \frac{0}{0}$$

$$\lim\limits_{x \to -2} \left(\frac{3x^2 + 2x - 8}{3x^2 + 16x + 20} \right) = \frac{0}{0} \Rightarrow \lim\limits_{x \to -2} \left(\frac{6x + 2}{6x + 16} \right) = \frac{-10}{4} = \frac{-5}{2}$$

Example: $\lim\limits_{x \to 0} \left(\dfrac{\cos x - 1}{x^2 - x} \right)$

$$\lim\limits_{x \to 0} \left(\frac{\cos x - 1}{x^2 - x} \right) = \frac{1 - 1}{0 - 0} = \frac{0}{0}$$

$$\lim\limits_{x \to 0} \left(\frac{-\sin x}{2x - 1} \right) = \frac{0}{-1} = 0$$

Example: $\lim\limits_{x \to 1} \left(\dfrac{\ln x - x + 1}{e^x - ex} \right)$

$$\lim\limits_{x \to 1} \left(\frac{\ln x - x + 1}{e^x - ex} \right) = \frac{0 - 1 + 1}{e - e} = \frac{0}{0}$$

$$\lim\limits_{x \to 1} \left(\frac{\frac{1}{x} - 1}{e^x - e} \right) = \frac{1 - 1}{e - e} = \frac{0}{0}$$

$$\lim\limits_{x \to 1} \left(\frac{\frac{-1}{x^2}}{e^x} \right) = \frac{-1}{e}$$

Example: $\lim\limits_{x \to \infty} \dfrac{x \ln x}{e^x - \sqrt{x}}$

$$\lim\limits_{x \to \infty} \frac{x \ln x}{e^x - 1} = \frac{\infty}{\infty}$$

$$\lim\limits_{x \to \infty} \frac{x \left(\frac{1}{x} \right) + \ln x}{e^x} = \frac{\infty}{\infty}$$

$$\lim\limits_{x \to \infty} \frac{\frac{1}{x}}{e^x} = 0$$

CHAPTER 5
PRACTICE PROBLEMS

(See solutions on page 224)

1. A 13-foot ladder leaning against a wall starts to slip in such a way that the foot of the ladder slips away from the wall at 2 in/sec. How fast is the top of the ladder slipping down the wall when the foot of the ladder is 12 inches from the wall?

2. A company has x boxes of produce available. The supply equation is given by $px - 10p + 20 = 3x$ where p is the price per box of produce in dollars per day and x is the number of boxes. If x is decreasing at 3 boxes per day, at what rate is the price changing when x is equal to 50 boxes?

3. If $f(x) = \dfrac{3x - 2}{2x + 1}$, use the tangent line at $x = 3$ to approximate $f'(x)$.

4. If $f(-1) = -3$ and $f'(-1) = 5$, approximate $f(-0.99)$ and $f(-1.01)$.

5. Find $\lim\limits_{x \to 0} \left(\dfrac{x \cos \pi x + \sin x}{x} \right)$

6. Find $\lim\limits_{x \to \infty} \left(\dfrac{4x^2 - 5x + 2}{e^{5x} + \ln x} \right)$

Analytical Applications of Differentiation

I. INTERMEDIATE-VALUE THEOREM

A. The Intermediate-Value Theorem states that a continuous function will take on all values between $f(a)$ and $f(b)$. The graph of a continuous function shown below illustrates the Intermediate-Value Theorem.

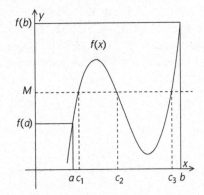

As the graph shows, if we pick any value, M, that is between the value of $f(a)$ and the value of $f(b)$ and draw a line straight out from this point, the line will intersect the graph in at least one point. Thus, somewhere between $x = a$ and $x = b$ the function will take on the value of M. Also, as the figure shows, the function may take on the value at more than one place. In this case, there are three values of c.

Note that the Intermediate-Value Theorem only states that the function will take on the value of M somewhere between a and b. It does not state what that value will be—just that it exists. The Intermediate-Value Theorem is normally used to show that there is a root of a function on a given interval.

Example 1: Show that $f(x) = 2x^3 - 8x^2 + 5x - 2$ has at least one root between $x = 1$ and $x = 4$.

Since a polynomial is continuous and $f(-1) = -17$ and $f(4) = 18$, by the Intermediate-Value Theorem there must be at least one value c on $(-1, 4)$ where $f(c) = 0$.

II. MEAN VALUE THEOREM

A. The Mean Value Theorem (MVT) states that if a function is continuous on $[a, b]$ and differentiable on (a, b) then there exists at least one x value, $x = c$, where $a < c < b$, such that $f'(c) = \dfrac{f(b) - f(a)}{b - a}$. In English, this says that a function is smooth (no breaks, corners, cusps, or vertical tangents), on an interval, then there must be at least one point in that interval at which the slope of the tangent line equals the slope of the secant line between the endpoints. Equivalently, there must be at least one point in the interval at which the tangent line is parallel to the secant between the endpoints.

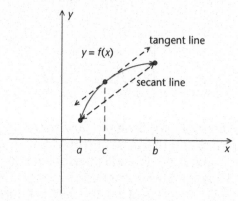

Example 1: Find the value of c guaranteed by the Mean Value Theorem for $f(x) = x^2$ on $[0, 3]$.

Since $f'(x) = 2x$, $f'(c) = 2c$. So, $f'(c) = \dfrac{f(b) - f(a)}{b - a} \rightarrow$

$2c = \dfrac{f(3) - f(0)}{3 - 0} = \dfrac{9}{3} = 3 \rightarrow 2c = 3 \rightarrow c = \dfrac{3}{2}$.

Example 2: Find the value of c guaranteed by the Mean Value Theorem for $f(x) = x^3 - 4x^2 - x + 4$ on $[-1, 2]$.

Since $f'(x) = 3x^2 - 8x - 1$, $f'(c) = 3c^2 - 8c - 1$. So,

$f'(c) = \dfrac{f(b) - f(a)}{b - a} \rightarrow 3c^2 - 8c - 1 = \dfrac{f(2) - f(-1)}{2 - (-1)} = \dfrac{-6}{3} = -2 \rightarrow$

$3c^2 - 8c + 1 = 0 \rightarrow$

$c \approx .131$, or $c \approx 2.535$. Final answer: $c \approx .131$ (Reject $c \approx 2.535$ since it is not in the interval given. Also, round off—always at the end of a problem—to at least three decimal places.)

III. ROLLE'S THEOREM

A. Rolle's Theorem states if a function is continuous on $[a, b]$ and differentiable on (a, b) and $f(a) = f(b)$, then there exists an x value, $x = c$, where $a < c < b$, such that $f'(c) = 0$. This is a simpler case of the Mean Value Theorem in which $f(a) = f(b)$. Clearly, if this is the case, the numerator of the fraction in the MVT becomes zero, thus $f'(c) = 0$.

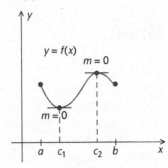

Example 1: Find the value of c guaranteed by Rolle's Theorem for $f(x) = x^3 - 4x^2 - x + 4$ on $[-1, 4]$.
$f(-1) = f(4) = 0$ and since $f'(x) = 3x^2 - 8x - 1$, $f'(c) = 3c^2 - 8c - 1$.
So, $f'(c) = 0 \rightarrow 3c^2 - 8c - 1 = 0 \rightarrow c \approx -.120$ or $c \approx 2.786$.

IV. WHEN DOES A FUNCTION *NOT* SATISFY EITHER OF THE ABOVE THEOREMS?

A. Sometimes you'll be asked to verify if a certain function satisfies either the MVT or Rolle's Theorem. All you need to do is to make sure it satisfies <u>all</u> of the hypotheses.

Example 1: $f(x) = \dfrac{1}{x}$ does not satisfy either theorem on an interval containing the origin because this function is not continuous (and hence, not differentiable) there.

Example 2: $g(x) = (x - 4)^{\frac{2}{3}}$ does not satisfy either theorem on any interval containing the point $(4, 0)$ because, though continuous there, it is not differentiable there (that is, $g'(4)$ does not exist, $g(x)$ has a cusp there).

Example 3: The function $f(x) = x^2$ does not satisfy Rolle's Theorem on $(0, 1)$ because $f(0) \neq f(1)$.

Keep in Mind...

➤ Don't confuse the Mean Value Theorem with Rolle's Theorem. Remember: for Rolle's Theorem you must set the function's derivative equal to zero, but for the Mean Value Theorem you must set the function's derivative equal to the slope of the secant between $x = a$ and $x = b$.

➤ Also, for both theorems, remember that the c value you are looking for is a number between a and b, but it cannot be a

or *b*. If there is a *c* value that falls outside of the given interval, reject it.

➤ Rolle's Theorem applies to a function on [*a*, *b*] only if $f(a) = f(b)$.

V. SKETCHING $f(x)$ GIVEN ITS EQUATION

A. Derivatives and intervals of increase and decrease
  1. If $f'(x) > 0$ on (*a*, *b*) then $f(x)$ is increasing on (*a*, *b*).
     Ex: $f(x) = x^2$ on (0, ∞). The converse is true as well.

  2. If $f'(x) < 0$ on (*a*, *b*), then $f(x)$ is decreasing on (*a*, *b*),
     Ex: $f(x) = x^2$ on (−∞, 0). The converse is true as well.

  3. If $f'(x) = 0$ at $x = a$ then $x = a$ is a candidate for the *x*-value of a max/min point. Ex: $f(x) = x^2$ at $x = 0$ there's a minimum point because $f'(0) = 0$ and f' changes sign from negative to positive here. Ex: At $x = 0$, $f(x) = x^3$ does not have a max or a min point because, even though $f'(0) = 0$, $f'(x)$ does not change sign here.

  4. If $f'(x)$ dne at $x = a$ and $f(a)$ exists, then $x = a$ is a candidate for the *x*-value of a max/min point.

     i. At $x = 0$, $f(x) = x^{\frac{2}{3}}$ has an absolute minimum point because $f'(0)$ dne, $f(0)$ exists, and $f'(x)$ changes sign from negative to positive.

     ii. At $x = 0$, $f(x) = x^{\frac{1}{3}}$ does not have a max or min point because, though $f'(0)$ does not exist and $f(0)$ exists, $f'(x)$ does not change sign here.

  5. Critical points of $f(x)$ are points in its domain at which $f'(x) = 0$ or $f'(x)$ does not exist.

B. Derivatives and concavity
  1. If $f''(x) > 0$ on (*a*, *b*), then $f(x)$ is concave up on (*a*, *b*).
     Ex: $f(x) = x^2$ on (−∞, ∞). The converse is true as well.

  2. If $f''(x) < 0$ on (*a*, *b*), then $f(x)$ is concave down on (*a*, *b*).
     Ex: $f(x) = -x^2$ on (−∞, ∞). The converse is true as well.

3. If $f''(x) = 0$, then $x = a$ is a candidate for the x-value of an *inflection point* (a point at which the concavity of $f(x)$ changes).

   i. $f(x) = x^3$ has an inflection point at $x = 0$ because $f''(0) = 0$ and $f''(x)$ changes sign from negative to positive. An inflection point occurs at $x = a$ if and only if $f(a)$ exists, $f''(a) = 0$ or does not exist, and $f''(x)$ changes signs at $x = a$.

C. Graphing a function requires finding its intercepts, relative and absolute extrema, and asymptotes.

   1. Intercepts—an x-intercept is a point at which a function intersects the x-axis, and hence, $y = 0$ here. A y-intercept is a point at which a function intersects the y-axis, and hence, $x = 0$ here. Not all functions have intercepts, for example, $y = \dfrac{1}{x}$. Some functions have an infinite number of x-intercepts, for example, $y = \sin(x)$. A function can have *at most* one y-intercept. If a graph has more than one y-intercept, it violates the definition of function because it would have more than one different y-value for $x = 0$. No calculus is necessary to find intercepts.

   2. Relative maximum/minimum points—a point on a function is a relative (or local) maximum point if and only if it is the highest point in its neighborhood. Think of it as the top of a mountain but not necessarily the highest mountain. A point on a function is a relative (or local) minimum point if and only if it is the lowest point in its neighborhood. Think of it as the bottom of a valley but not necessarily the lowest valley. Relative (or local) extrema (that is, maximum or minimum points) occur at interior points of a function, not at endpoints. Not all functions have relative extrema, for example, $y = \dfrac{1}{x}$. The relative extrema occur at points where the first derivative is either zero or nonexistent *and* the function is defined. In particular, if the minimum of a function occurs at an interior point of the function, $x = a$, the derivative is negative to the left of a and positive to the right of a—that is, the function must change from decreasing to increasing at $x = a$. If the maximum of a function occurs at an

interior point of the function, $x = b$, the derivative is positive to the left of b and negative to the right of b—that is, the function must change from increasing to decreasing at $x = b$.

3. Absolute maximum/minimum points—a point on a function is an absolute maximum point if and only if it is the highest point. Think of it as the top of the highest mountain. A point on a function is an absolute minimum point if and only if it is the lowest point. Think of it as the bottom of the lowest valley. Absolute extrema can occur at interior points or at endpoints of a function. The absolute extrema occur at points where the first derivative is either zero or nonexistent *and* the function is defined—or at endpoints of the function.

4. Critical points—these are points in the domain of a function at which the derivative is either equal to zero or does not exist. These are generally found when looking for max/min points.

5. Asymptotes—refer to Chapters 3 and 6.

6. The following functions are defined on the interval shown in the graph.

i.

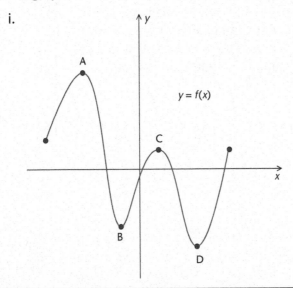

Point A is a relative maximum point and the absolute maximum point.
Point C is a relative maximum point.
Point B is a relative minimum point.
Point D is a relative minimum point and the absolute minimum point.

ii.

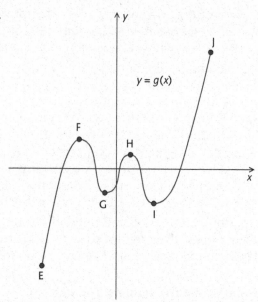

y = g(x)

Point J is the absolute maximum point,
but not a relative maximum.
Points F and H are relative maxima.
Points G and I are relative minima.
Point E is the absolute minimum point,
but not a relative minimum.

iii.

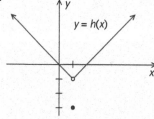

y = h(x)

The absolute minimum point of y = h(x) is (1, −3). At x = 1,
h(x) is defined but its derivative does not exist.

Test Tip

A point on a function consists of both the x and y values of the point, so when asked to find a minimum/maximum point, find both the x and y values of it. When asked to find the minimum/maximum value of a function, report only the y-value.

iv. Find the absolute minimum point of $f(x) = xe^x$.

Step 1: Find $f'(x)$. $f'(x) = xe^x + e^x$

Step 2: Set $f'(x) = 0$ and also check for points where $f'(x)$ does not exist. $f'(x) = xe^x + e^x = 0 \rightarrow e^x(x+1) = 0 \rightarrow x = -1$ (e^x is positive for all values of x). This function has no points of nondifferentiability.

Step 3: Check to see if there is a max or min at $x = -1$ by making a sign analysis chart for $f'(x)$. Make sure to include the x-values found in step 2, then a value from the right and a value from the left of those x-values.

$f'(x)$	negative	**0**	positive
x	-2	**-1**	0

Since $f(-1)$ exists, $f'(-1) = 0$ and $f'(x)$ changes from negative to positive at $x = -1$, we conclude that the absolute minimum point of $f(x)$ occurs at $x = -1$.

Step 4: Since asked to find the absolute minimum *point* of $f(x)$, find the value of $f(-1) = (-1)e^{-1} = -\dfrac{1}{e}$. Final answer: $\left(-1, -\dfrac{1}{e}\right)$. The graph of $f(x) = xe^x$:

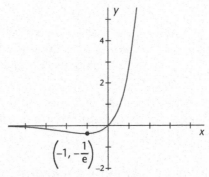

v. Find the absolute maximum and minimum values of $g(x) = -x^4 - 2x^3$ on $[-2, 2]$

Step 1: Find $g'(x)$. $g'(x) = -4x^3 - 6x^2$

Step 2: Set $g'(x) = 0$ and look for points of nondifferentiability. $g'(x) = -4x^3 - 6x^2 = 0 \rightarrow$ $-2x^2(2x + 3) = 0 \rightarrow x = 0$ or $x = -\dfrac{3}{2}$. $g(x)$ is a polynomial so it has no points of nondifferentiability.

Step 3: Make a sign analysis chart for $g'(x)$ making sure to include the endpoints since they are candidates for absolute extrema.

$g'(x)$	positive	**0**	negative	**0**	negative
x	-2	$-\dfrac{3}{2}$	-1	0	2

Since $g'(x)$ changes from positive to negative at $x = -\dfrac{3}{2}$, and $g\left(-\dfrac{3}{2}\right)$ exists, $g(x)$ must have a maximum here.

Evaluate the original function at the critical points, $\left(x = -\dfrac{3}{2}, x = 0\right)$ and at the endpoints to find absolute extrema: $g\left(-\dfrac{3}{2}\right) = -\left(-\dfrac{3}{2}\right)^4 - 2\left(-\dfrac{3}{2}\right)^3 =$ 1.6875; $g(0) = 0$; $g(-2) = 0$; $g(2) = -32$. The highest y-value is $y = 1.6875$ so this is the absolute maximum value of $g(x)$. The lowest y-value is $y = -32$ so this is the absolute minimum of $g(x)$. Note that you were *not* asked to find the absolute maximum and minimum *points* of $g(x)$ but only the absolute maximum and minimum values of $g(x)$.

vi. Find the critical points of $f(x) = \sqrt{x}$ and characterize them as absolute maximum or absolute minimum points.

Step 1: Find y'. $y' = \dfrac{1}{2\sqrt{x}}$

Step 2: In this case, note that $f'(x)$ does not equal zero for any x-value and $f'(x)$ does not exist at $x = 0$. Since $f(0)$ is defined, there is a critical point at $x = 0$ and this critical point is a candidate for absolute max/min.

Step 3: Make a sign analysis chart for $f'(x)$.

$f'(x)$	**dne**	positive
x	**0**	2

Note that we cannot include negative values in the sign analysis chart since these are not included in the domain of $f(x)$.

From this chart we conclude that the function, $y = \sqrt{x}$, increases without bound since its derivative is positive on $(0, \infty)$.

Step 4: Since y is increasing on $x > 0$, the lowest y value must occur at $x = 0$ and $y\big|_{x=0} = 0$. According to the sign analysis chart for $f'(x)$, $f(x)$ has only one absolute minimum point at $(0, 0)$, its left endpoint.

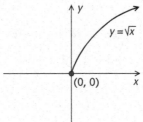

vii. Given $f(x) = \dfrac{x^2 - 1}{x^2 - 4}$ find its intercepts, asymptotes, intervals of increase and decrease, relative max/min points, absolute max/min points, concavity intervals, points of inflection and graph it.

Intercepts:

x-intercept, set $y = 0$: $\dfrac{x^2 - 1}{x^2 - 4} = 0 \rightarrow x^2 - 1 = 0 \rightarrow$

$x = \pm 1$ (Note that the denominator does not equal zero at $x = \pm 1$.)

y-intercept: set $x = 0$: $\dfrac{0^2 - 1}{0^2 - 4} = \dfrac{1}{4} \rightarrow y = \dfrac{1}{4}$

Vertical asymptotes: $x^2 - 4 = 0 \rightarrow x = \pm 2$. Note that only the denominator equals zero at $x = \pm 2$, not the numerator.

<u>Horizontal asymptotes</u>: $\lim\limits_{x\to\infty}\dfrac{x^2-1}{x^2-4}=1$ and $\lim\limits_{x\to-\infty}\dfrac{x^2-1}{x^2-4}=1\to$

Horizontal asymptote: $y=1$.

Note: If the limits were equal to different numbers, then the function would have two different horizontal asymptotes. If one limit were equal to a number, $y=b$, and the other to plus or minus infinity, $y=b$ would be the only horizontal asymptote. If both limits were equal to plus or minus infinity, then the function would not have any horizontal asymptotes.

<u>Intervals of increase/decrease and max/min points</u>

$f'(x)=-\dfrac{6x}{(x^2-4)^2}=0$ at $x=0$. $f'(x)=-\dfrac{6x}{(x^2-4)^2}$ does not

exist at $x=\pm2$

$f'(x)$	positive	**dne**	positive	**0**	negative	**dne**	negative
x	-3	-2	-1	0	1	2	3

According to the sign analysis chart for $f'(x)$, $f(x)$ is increasing on $(-\infty, -2) \cup (-2, 0)$ and decreasing on $(0, 2) \cup (2, \infty)$. Relative maximum point occurs at $\left(0, \dfrac{1}{4}\right)$.

<u>Intervals of concavity and inflection points</u>

$f''(x)=\dfrac{18x^2+24}{\left(x^2-4\right)^3}\neq0$ since the numerator is positive for

all x values. $f''(x)=\dfrac{18x^2+24}{\left(x^2-4\right)^3}$ does not exist at $x=\pm2$

but since $f(\pm2)$ does not exist, there are no inflection points for $f(x)$.

$f''(x)$	positive	**dne**	negative	**dne**	positive
x	-3	-2	0	2	3

According to the sign analysis chart for $f''(x)$, $f(x)$ is concave up on $(-\infty, -2) \cup (2, \infty)$ and concave down on $(-2, 2)$.

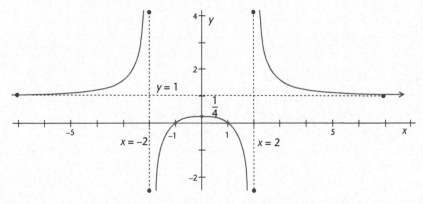

VI. SKETCHING $f(x)$ GIVEN THE GRAPH OF $f'(x)$

A. When given the graph of $f'(x)$ create a sign analysis chart for $f'(x)$ and then draw $f(x)$ based on it and any other information given. Analyzing the slopes of the graph of $f'(x)$ also helps you to find the concavity intervals of $f(x)$.

1. Given the graph of $f'(x)$ below, and $f(-4) = f(5) = -1$ and $f(0) = 3$, sketch the graph of $f(x)$.

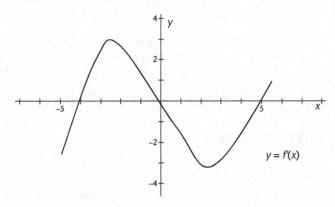

x	$x < -4$	-4	$-4 < x < -2.5$	-2.5	$-2.5 < x < 0$	0	$0 < x < 2.5$	2.5	$2.5 < x < 5$	5	$x > 5$
$f'(x)$	–	zero	+	max	+	zero	–	min	–	zero	+
$f''(x)$	+	+	+	zero	–	–	–	zero	+	+	+
$f(x)$	decreasing concave up	minimum concave up	increasing concave up	inflection point	increasing concave down	maximum concave down	decreasing concave down	inflection point	decreasing concave up	minimum concave up	increasing concave up

According to the sign analysis charts for $f'(x)$ and $f''(x)$, the graph of $f(x)$ looks like the following:

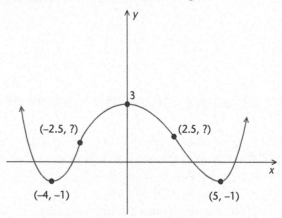

Note 1: The *y*-value of the inflection point is not known and it is not necessary for sketching the original function.

Note 2: The graph of the derivative looks like a cubic function with a positive leading coefficient. Therefore, expect the graph of the original, that of $f(x)$, to look like a quartic function with a positive leading coefficient.

VII. ABSOLUTE EXTREMA

A. The extreme value theorem is an existence theorem. It states that if a function *f* is continuous on the closed interval [*a*, *b*], then *f* must attain an absolute maximum value and absolute minimum value at least once. The theorem seems obvious but is not easy to prove. It is extremely relevant to everyday lives. Sitting in your classroom, there is a student who is taller than everyone else, as well as a student who is shorter than everyone else. There is a student with the highest IQ and one with the lowest.

1. Here's how to find an absolute maximum and absolute minimum on a closed interval [*a*, *b*]. Absolute extrema (sometimes called global extrema) can also be relative extrema (local extrema) and vice versa. For instance, in the graph, $f(x) = x^2 - 2x - 1$ on the interval [3, –4], the point (1, –2) is the absolute minimum as well as a relative minimum. The point (–3, 14) is the absolute maximum but not a relative maximum.

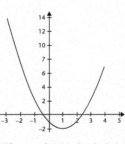

2. To find absolute extrema on an interval, look for candidates for the absolute maximum and absolute minimum of *f* on [*a*, *b*] by finding *x*-values and determining which gives the largest and smallest value of *f*.
   i.   Find critical values of *f* on (*a*, *b*) by setting the numerator and denominator of *f*'(0) and solving.
   ii. Evaluate *f* at each critical value of *f* on (*a*, *b*).
   iii. Evaluate *f* at each endpoint of [*a*, *b*]—that is find *f*(*a*) and *f*(*b*). Extrema cannot occur at open endpoints.
   iv. The largest of these is the absolute maximum. The smallest of these is the absolute minimum.
   v.   Remember to discern the difference *where* the absolute maximum or absolute minimum occurs as opposed to *what* the absolute maximum or absolute minimum is. *Where* is the *x*-value. *What* is the *y*-value.

3. Find the absolute maximum and minimum values of the following functions on the given interval.
   1) $f(x) = 3x^2 - 24x - 1$ [–1, 5]

   $f'(x) = 6x - 24 = 0$
   $6x = 24 \Rightarrow x = 4$

x	*f*(*x*)
4	–49 — Abs min
–1	26 — Abs max
5	–45

2) $f(x) = \dfrac{x^2}{x^2+3}$ $[-1, 1]$

$f'(x) = \dfrac{2x(x^2+3) - x^2(2x)}{(x^2+3)^2}$

$\dfrac{6x}{(x^2+3)^2} = 0 \Rightarrow x = 0$

x	$f(x)$
0	0 — Abs min
−1	1/4 — Abs max
1	1/4 — Abs max

 ## VIII. OPTIMIZATION

A. Optimizing a quantity means to find its maximum or minimum value. For instance, one could find the maximum profit or the minimum loss in a business situation. These are word problems so they must be read carefully. The steps to solve an optimization problem are:

(a) Create a legend that includes the given information and the variable that you are looking for.

(b) Write down the function that needs to be optimized in terms of one variable. If there are two variables, then write a secondary equation that links them.

(c) Take the derivative of the function in part b, set it equal to zero, and solve. Justify, using a sign analysis chart, that you found a maximum or a minimum, as the case may be.

(d) Double-check that you found the answer to the question being asked.

 **Test Tip**

Make sure to include correct units! If the optimization problem is in the free-response section, write your answer in a complete sentence.

Example: Find the radius of the largest cylinder that can be inscribed in a cone of radius 3 in. and height 5 in.

(a) Create legend: r_{cone} = 3 in., h_{cone} = 5 in., $r_{cylinder}$ = ? such that $V_{cylinder}$ is maximum? (Optimize $V_{cylinder}$)

(b) The function to be optimized: $V_{cylinder} = \pi r^2_{cylinder} h_{cylinder}$. Rewrite $h_{cylinder}$ in terms of $r_{cylinder}$ so the function contains only one variable and can be more easily differentiated. In this case, use the fact that the ratios of corresponding sides of similar triangles are equal. That is,

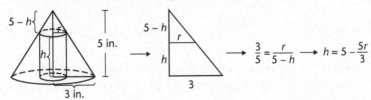

So, $V_{cylinder} = \pi r^2_{cylinder}\left(5 - \dfrac{5r_{cylinder}}{3}\right) = 5\pi r^2_{cylinder} - \dfrac{5}{3}\pi r^3_{cylinder}$

(c) Take the derivative of the function in part (b) and set it equal to zero: $\dfrac{dV_{cylinder}}{dr} = 10\pi r_{cylinder} - 5\pi r^2_{cylinder} = 0 \rightarrow$

$r_{cylinder} = 2\,\text{in.}$ (disregard $r_{cylinder} = 0$). Sign analysis chart for $V'_{cylinder}$ shows that when $r_{cylinder} = 2\,\text{in.}$, $V_{cylinder}$ is a maximum (largest).

$r_{cylinder}$	1	2	3
$\dfrac{dV_{cylinder}}{dr}$	positive	0	negative

(d) The radius of the largest cylinder that can be inscribed in a cone of radius 3 in. and height 5 in. is 2 in.

CHAPTER 6
PRACTICE PROBLEMS

(See solutions on page 226)

1. Find the *c* value guaranteed by the Mean Value Theorem for $f(x) = \dfrac{1}{x-1}$ on [2, 4].

2. Find the *c* value(s) guaranteed by Rolle's Theorem for $y = 2\cos(3x)$ on $[-\pi, \pi]$.

3. Does $y = \ln(x)$ satisfy the Mean Value Theorem on $[1, e]$? If yes, find *c*. If not, explain why not.

4. Does $y = \ln(x)$ satisfy Rolle's Theorem on any interval? Explain.

5. If $f(x)$ is a polynomial function passing through the following points $(1.5, -1)$, $(2.3, 2)$, $(3, 0)$, $(4.2, 8)$, $(5, -4)$, what is the minimum number of roots it has on $[0, 5]$?

6. Find the critical points, inflection points, the absolute minimum value of *y*, and relative maximum points of $y = x^4 - 3x^2 + 2$.

7. Sketch the graph of $f(x)$ if the graph of $f'(x)$ is given below:

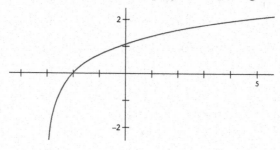

PART IV
INTEGRATION

Integration and Accumulation of Change

I. TYPES OF INTEGRALS

A. *Indefinite integrals* have no limits, $\int f(x)dx$. This represents the antiderivative of $f(x)$. $f(x)$, the expression being integrated, is called the integrand. That is, if $\int f(x)dx = F(x) + C$, then

$F'(x) = f(x)$. When taking an antiderivative of a function, don't forget the $+ C$ where C is a constant of integration. For instance, $\int 2xdx = x^2 + C$. (The constant C is necessary because the antiderivative of $f(x) = 2x$ could be $F(x) = x^2$ or $F(x) = x^2 + 1$ or $F(x) = x^2 - 2$, and so on.) Sometimes, you are given an initial condition that allows you to find the value of C. For instance, find the antiderivative, $F(x)$, of $f(x) = 2x$, given that $F(0) = 1$. Then,

$F(x) = \int 2xdx = x^2 + C \rightarrow F(0) = (0)^2 + C = 1 \rightarrow C = 1 \rightarrow F(x) = x^2 + 1$.

Another way of posing this question is: Find y if $\dfrac{dy}{dx} = 2x$ and

$y\big|_{x=0} = 1$. The equation $\dfrac{dy}{dx} = 2x$ is called a differential equation (covered in Chapter 8) because it contains a derivative.

B. *Definite integrals* have limits $x = a$ and $x = b$, $\int_a^b f(x)dx$. If $f(x)$ is

continuous on $[a, b]$ and $F'(x) = f(x)$, then $\int_a^b f(x)dx = F(b) - F(a)$

(The First Fundamental Theorem of Calculus.)

1. A definite integral value could be positive, negative, zero or infinity. When used to find area, the definite integral must have a positive value.

i. If $f(x) > 0$ on $[a, b]$, then $\int_a^b f(x)dx > 0$ and geometrically

it represents the area between the graph of $f(x)$ and the

x-axis on the interval $[a, b]$. For example, $\int_0^3 2xdx = x^2 \Big]_0^3 = 9$

square units. Note that this could also have been solved geometrically because the area in question is that of a right triangle with a base of 3 units and a height of 6 units.

$(A_\Delta = \dfrac{1}{2}bh = \dfrac{1}{2}3(6) = 9 \text{ units}^2)$

Solving an area problem geometrically is really helpful when the question involves the integral of a piecewise

linear function, for instance, $\int_{-1}^3 |x|dx$. This represents the area

between the function $f(x) = |x|$ and the x-axis between $x = -1$ and $x = 3$. Noticing that this area is that of two right

triangles, we have: $\int_{-1}^3 |x|dx = \dfrac{1}{2}1(1) + \dfrac{1}{2}3(3) = 5 \text{ units}^2$.

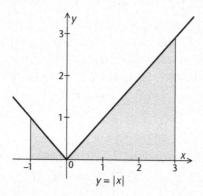

$y = |x|$

Remember that an absolute-value function is always made up of two pieces which you most often must consider separately because each piece is defined on a different interval. For

instance, recall that $|x| = \begin{cases} x, & x \geq 0 \\ -x, & x < 0 \end{cases}$. So, algebraically,

$\int_{-1}^3 |x|dx = \int_{-1}^0 -xdx + \int_0^3 xdx = -\dfrac{x^2}{2}\Big]_{-1}^0 + \dfrac{x^2}{2}\Big]_0^3 = \left(0 + \dfrac{1}{2}\right) + \left(\dfrac{9}{2} - 0\right) =$

5 units². This would take much more time, especially when the functions get more complicated.

ii. If $f(x) < 0$ on $[a, b]$ then $\int_a^b f(x)dx < 0$ and the area between the graph of $f(x)$ and the x-axis on the interval $[a, b]$ is represented either by $\int_a^b |f(x)|dx$ or by $\left|\int_a^b f(x)dx\right|$. For instance, $\int_{-1}^1 (x^2 - 1)dx$. If you are simply asked to evaluate the integral, do so. That is,

$$\int_{-1}^1 (x^2 - 1)dx = \frac{x^3}{3} - x\Big]_{-1}^1 = \left(\frac{1}{3} - 1\right) - \left(\frac{-1}{3} - (-1)\right) = -\frac{4}{3}.$$

However, if the question asks for area, use the absolute value since area is always positive. That is, write either

$$\int_{-1}^1 |(x^2 - 1)| dx = -\int_{-1}^1 (x^2 - 1)dx = -\left(\frac{x^3}{3} - x\right)\Big]_{-1}^1 =$$

$$-\left[\left(\frac{1}{3} - 1\right) - \left(\frac{-1}{3} - (-1)\right)\right] = \frac{4}{3} \text{ square units or,}$$

$$\left|\int_{-1}^1 (x^2 - 1)dx\right| = \frac{x^3}{3} - x\Big]_{-1}^1 = \left|\left(\frac{1}{3} - 1\right) - \left(\frac{-1}{3} - (-1)\right)\right| = \frac{4}{3} \text{ square}$$

units.

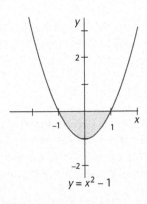

$y = x^2 - 1$

iii. If $f(x)$ is positive for some values of x and negative for other values of x on $[a, b]$, then the area between the graph of $f(x)$ and the x-axis on the interval $[a, b]$

is represented by $\int_a^b |f(x)|\,dx$. For example, the area between $f(x) = x^2 - 1$ and the x-axis on $[0, 2]$ is given by

$$\int_0^2 |x^2 - 1|\,dx = \int_0^1 -(x^2 - 1)\,dx + \int_1^2 (x^2 - 1)\,dx =$$

$$-\frac{x^3}{3} + x\bigg]_0^1 + \frac{x^3}{3} - x\bigg]_1^2 = \left(-\frac{1}{3} + 1\right) + \left[\left(\frac{8}{3} - 2\right) - \left(\frac{1}{3} - 1\right)\right] = 2\text{ units.}^2$$

If you simply take the integral,

$$\int_0^2 (x^2 - 1)\,dx = \frac{x^3}{3} - x\bigg]_0^2 = \frac{8}{3} - 2 = \frac{2}{3}, \text{ the answer does not}$$

represent the area between the function and the x-axis, it represents the difference between the area above and the area below the x-axis.

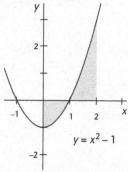

$$y = x^2 - 1$$

Note: While definite integrals represent area measured in square units, unless the units are specified, we simply use a number to represent the area.

C. The Accumulation Function

Definite integrals refer to the area under a curve. The limits of integration indicate the x-values where the area process starts and the area process ends. We now make the upper limit of integration a variable, usually x. Our integral structure is now changed to: $g(x) = \int_a^x f(t)\,dt$. This says we will start finding the area process at some constant a and end the area process at some variable x. Since this is a function of the variable x, use $f(t)\,dt$ rather than $f(x)\,dx$ to prevent confusion. This is sometimes called the accumulation function because we are accumulating area under a curve based on the value of x.

1. Following is the graph of $y = f(x)$ on the domain $[-4, 5]$, made up of lines. Let $g(x) = \int_{0}^{x} f(t)\, dt$.

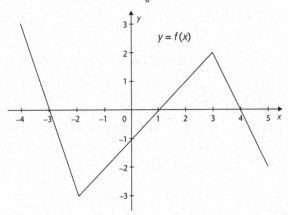

$g(4) = \int_{0}^{4} f(t)\, dt$ is the accumulated area under the f curve starting at 0 and ending at 4. This is $-0.5 + 3 = 2.5$.

$g(-3) = \int_{0}^{-3} f(t)\, dt$ is the accumulated area under the f curve starting at 0 and ending at -3. It follows that

$$\int_{0}^{-3} f(t)\, dt = -\int_{-3}^{0} f(t)\, dt = 5.5.$$

Since by the Second Fundamental Theorem of Calculus,

$$\frac{d}{dx} \int_{a}^{x} f(t)\, dt = f(x), \text{ then } g'(4) = f(4) = 0.$$

To find the maximum value of g, we use this fact. Since $g'(x) = f(x)$, there are relative maxima at $x = -3$ and $x = 4$. Since $g(x) = 2.5$ and $g(-3) = 5.5$, the maximum value of g is 5.5.

D. *Improper Integrals** have one or both limits equal to either positive or negative infinity or are discontinuous at some value of x between the two limits of integration.

* This is a BC-only topic.

1. $\int_{1}^{\infty} \frac{1}{x^2} dx = \lim_{t \to \infty} \int_{1}^{t} \frac{1}{x^2} dx = \lim_{t \to \infty} \left(-\frac{1}{x} \right) \Big|_{1}^{t} = 0 - (-1) = 1$. If the answer is a constant, we say that the integral converges. If the answer is $\pm\infty$, we say that the integral diverges.

2. $\int_{-\infty}^{\infty} e^x dx = \left[e^x \right]_{-\infty}^{\infty} = e^{\infty} - e^{-\infty} = e^{\infty} - 1 = e^{\infty}$. The integral diverges.

3. $\int_{-1}^{1} \frac{1}{x} dx$. This is an improper integral because $y = \frac{1}{x}$ is

   discontinuous at $x = 0$. Thus, $\int_{-1}^{1} \frac{1}{x} dx = \int_{-1}^{0} \frac{1}{x} dx + \int_{0}^{1} \frac{1}{x} dx =$

   $\left[\ln|x| \right]_{-1}^{0} + \left[\ln|x| \right]_{0}^{1}$. This is $\ln 0 - \ln 1 + \ln 1 - \ln 0 = \infty - 0 + 0 - \infty$.
   Do not make the mistake of canceling the infinities. Once any part of the calculation is infinity, the integral diverges.

II. PROPERTIES OF DEFINITE INTEGRALS

1. $\int_{a}^{b} kf(x)dx = k\int_{a}^{b} f(x)dx$, for any constant k

2. $\int_{a}^{a} f(x)dx = 0$

3. $\int_{a}^{b} f(x)dx = -\int_{b}^{a} f(x)dx$

4. $\int_{a}^{b} f(x)dx + \int_{b}^{c} f(x)dx = \int_{a}^{c} f(x)dx$

5. $\int_{a}^{b} [f(x) \pm g(x)]dx = \int_{a}^{b} f(x)dx \pm \int_{a}^{b} g(x)dx$

6. If $f(x) \le g(x)$ on $[a, b]$, then $\int_{a}^{b} f(x)dx \le \int_{a}^{b} g(x)dx$

7. If $f(x)$ is even on $[-a, a]$, then $\int_{-a}^{a} f(x)dx = 2\int_{0}^{a} f(x)dx$

8. If $f(x)$ is odd on $[-a, a]$, then $\displaystyle\int_{-a}^{a} f(x)dx = 0$

9. $\displaystyle\int_{a}^{b} f'(x)dx = f(b) - f(a)$

III. THEOREMS

A. <u>The First Fundamental Theorem of Calculus</u> states that if $f(x)$ is continuous on $[a, b]$ and $F'(x) = f(x)$, then $\displaystyle\int_{a}^{b} f(x)dx = F(b) - F(a)$ or $F(b) = F(a) + \displaystyle\int_{a}^{b} f(x)dx$.

B. <u>The Second Fundamental Theorem of Calculus</u> states that if $f(x)$ is continuous on $[a, b]$, then $\dfrac{d}{dx}\displaystyle\int_{a}^{x} f(t)dt = f(x)$. In general, $\dfrac{d}{dx}\displaystyle\int_{a}^{g(x)} f(t)dt = f(g(x))(g'(x))$.

Keep in Mind...

➤ Don't forget to add the constant, C, when finding an indefinite integral.

➤ Whenever possible, try to work backwards to find the antiderivative of a function. It might save time.

➤ Don't confuse $\left|\displaystyle\int_{a}^{b} f(x)dx\right|$ with $\displaystyle\int_{a}^{b} |f(x)|dx$. They are the same only in intervals $[a, b]$ such that $f(x)$ is always positive or $f(x)$ is always negative.

➤ Remember that an absolute value function is a piecewise function. Thus, when integrating an absolute value function, integrate each piece separately or do the problem graphically.

IV. **RIEMANN SUMS (LRAM, RRAM, MRAM)** are used to approximate the area between a function and the *x*-axis by slicing the area into thin vertical rectangles. (Riemann sums are sometimes used to approximate the area between a function/relation and the *y*-axis.)

Note: For all examples below, we will use Riemann sums with 5 rectangles. The method works with any number of rectangles with more giving more accurate answers, and of course, generating more work.

A. LRAM—Left Rectangle Approximation Method. To approximate the area between a function, $f(x)$, and the *x*-axis on $[a, b]$, slice the area into vertical rectangular strips each of width Δx (the value of Δx will be given in the problem). Starting on the left, create rectangles and add up all their areas. For instance, below is the graph of $f(x)$ on $[a, b]$. To approximate this area using LRAM, create rectangles as shown below:

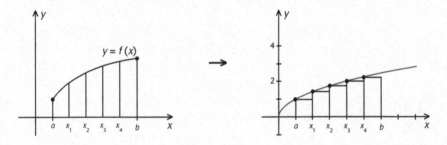

Then, the area between $f(x)$ and the *x*-axis on $[a, b]$ can be approximated by: $A \approx \Delta x(f(a) + f(x_1) + f(x_2) + f(x_3) + f(x_4))$.

This method is called the Left Rectangle Approximation Method because the upper left corner of each rectangle is on the curve. Note that in this case the approximation is an underestimation of the area since the area between the curve and the rectangles is left out.

B. RRAM—Right Rectangle Approximation Method. Using the function $f(x)$, seen above, create rectangles starting on the right such that the upper right corner of each rectangle is on the curve. This area then is represented by $A \approx \Delta x(f(x_1) + f(x_2) + f(x_3) + f(x_4) + f(b))$. Note that in this case the approximation is an *overestimation* of the actual area since the rectangles include more than just the area below the curve.

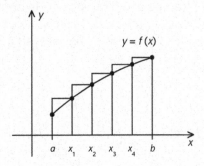

C. MRAM—<u>M</u>idpoint <u>R</u>ectangle <u>A</u>pproximation <u>M</u>ethod. Using the function $f(x)$, above, create rectangles such that the height of each rectangle is in the middle and the midpoint of the upper width of each rectangle is on the curve. This area is represented by:

$$A \approx \Delta x \left(f\left(\frac{a + x_1}{2}\right) + f\left(\frac{x_1 + x_2}{2}\right) + f\left(\frac{x_2 + x_3}{2}\right) + f\left(\frac{x_3 + x_4}{2}\right) + f\left(\frac{x_4 + b}{2}\right) \right).$$

In this case we're not exactly sure if this approximation is an underestimate or an overestimate because the rectangles are below as well as above the curve. However, it is clear that this method more closely approximates the actual area.

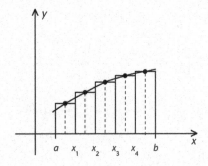

D. Trapezoid Rule—Given the function $f(x)$, seen earlier, connect the top endpoints of the vertical line segments, thus creating trapezoids. Add up the areas of all the trapezoids. This is an approximation of the area between the curve and the x-axis. Since the area of a trapezoid is $A = \frac{1}{2}h(b_1 + b_2)$, the sum of the areas of the trapezoids below is $A = \dfrac{1}{2}\Delta x\,(f(a) + 2f(x_1) + 2f(x_2) + 2f(x_3) + 2f(x_4) + f(b))$.

Note that the first base and the last base do not repeat, so they are not doubled. However, the inside bases are counted twice because adjacent trapezoids share a base. Also note that $h = \Delta x$.

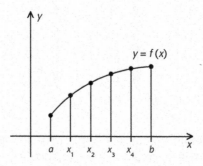

An example of all the above methods: approximate the area between the graph of $f(x) = \sqrt{x}$ and the x-axis on the interval $[1, 4]$ using $\Delta x = 1$:

1. LRAM: $A \approx 1(\sqrt{1} + \sqrt{2} + \sqrt{3}) = 4.14626437$

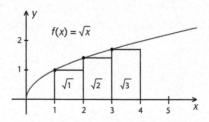

2. RRAM: $A \approx 1(\sqrt{2} + \sqrt{3} + \sqrt{4}) = 5.14626437$

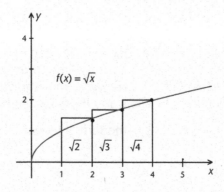

3. MRAM: $A \approx 1(\sqrt{1.5} + \sqrt{2.5} + \sqrt{3.5}) = 4.676712395$

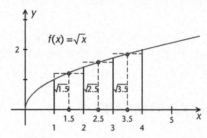

4. Trapezoid Rule: $A \approx \dfrac{1}{2}(1)(\sqrt{1} + 2\sqrt{2} + 2\sqrt{3} + \sqrt{4}) = 4.64626437$

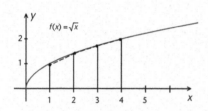

Trapezoid 1: $\dfrac{1}{2}\left(\sqrt{1} + \sqrt{2}\right)$

Trapezoid 2: $\dfrac{1}{2}\left(\sqrt{2} + \sqrt{3}\right)$

Trapezoid 3: $\dfrac{1}{2}\left(\sqrt{3} + \sqrt{4}\right)$

Realize that the results of the trapezoid rule is simply the average of the left Riemann sum and the right Riemann sum:

$$\frac{\left(1 + \sqrt{2} + \sqrt{3}\right) + \left(\sqrt{2} + \sqrt{3} + \sqrt{4}\right)}{2} = \frac{1}{2}\left(1 + 2\sqrt{2} + 2\sqrt{3} + \sqrt{4}\right)$$

This relationship is always true, no matter the function and the interval. So if you have calculated LRAM and RRAM, then you can average the results to get the trapezoid approximation.

5. The actual area: $A = \displaystyle\int_{1}^{4} x\,dx = \dfrac{2}{3}x^{3/2}\Big|_{1}^{4} = \dfrac{2}{3}(8-1) = \dfrac{14}{3} = 4.\bar{6}$

Keep in Mind...

➤ Whichever estimation you use, if it calculates more than the given area, it is an overestimation. If it calculates less than the given area, it's an underestimation.

➤ When asked to approximate the area between a curve and an axis using a Riemann sum, draw the diagram. It always helps.

➤ Generally, the MRAM and TRAP methods give better approximations than the LRAM and RRAM methods.

V. INTEGRATION TECHNIQUES

A. *U-substitution* is used to rewrite the integrand so that it is easily integrable. This method is used when the integrand is of the form $f(g(x))g'(x)$ where $g'(x)$ can be off by a constant factor. This is the opposite of the chain rule for derivatives.

1. To use this method for $\int f(g(x))g'(x)dx$, let $u = g(x)$. Then, $du = g'(x)dx$ and $\int f(g(x))g'(x)dx = \int f(u)du$.

   i. For instance, for $\int (6x - 2)\sqrt{3x^2 - 2x + 5}\,dx$,

   let $u = 3x^2 - 2x + 5$, $du = (6x - 2)dx$ and

   $$\int (6x - 2)\sqrt{3x^2 - 2x + 5}\,dx = \int \sqrt{u}\,du$$
   $$= \frac{2}{3}u^{\frac{3}{2}} + C = \frac{2}{3}(3x^2 - 2x + 5)^{\frac{3}{2}} + C.$$

 Test Tip

Always rewrite the problem using the original variable unless otherwise directed.

   ii. For $\int \dfrac{3x}{e^{5x^2}}\,dx$ let $u = 5x^2$ and $du = 10x\,dx \rightarrow \dfrac{1}{10}du = x\,dx$. So,

   $$\int \frac{3x}{e^{5x^2}}\,dx = 3\int \frac{x}{e^{5x^2}}\,dx = 3\int \frac{x\,dx}{e^{5x^2}} = 3\int \frac{\frac{1}{10}du}{e^u} = \frac{3}{10}\int \frac{1}{e^u}\,du =$$
   $$\frac{3}{10}\int e^{-u}\,du = -\frac{3}{10}e^{-u} + C = -\frac{3}{10}e^{-(5x^2)} + C.$$

iii. Find $\int \sin 5x\,dx$ $u = 5x$

$\dfrac{1}{5}\int \sin u\,du = \dfrac{1}{5}(-\cos u)$ $du = 5\,dx$

$\dfrac{-\cos x}{5} + C.$

B. *Integration by parts** is used when the integrand is a product of unrelated functions, of the form $f(x)g'(x)$. Let $u = f(x)$ and $v = g(x)$. Then, $\int u\,dv = uv - \int v\,du$. This is the opposite of the product rule for derivatives.

Test Tip

There are problems in which this method might be used more than once. To decide which of the two functions to let equal u and which to let equal dv can be tricky, but you'll know when you've gone down the wrong path. Instead of becoming simpler, the problem becomes more difficult.

i. For example, $\int xe^x\,dx$. Let $u = x$ and $dv = e^x\,dx$. Then, $du = dx$ and $v = e^x$. Thus, $\int xe^x\,dx = xe^x - \int e^x\,dx = xe^x - e^x + C.$

ii. When solving $\int x^2 \sin(x)\,dx$, let $u = x^2$ and $dv = \sin(x)\,dx$. Then, $du = 2x\,dx$ and $v = -\cos(x)$. Thus, $\int x^2 \sin(x)\,dx = -x^2 \cos(x) + 2\int x\cos(x)\,dx$. Here we need to integrate by parts again. So, to integrate $\int x\cos(x)\,dx$ let $u = x$ and $dv = \cos(x)\,dx$. Then $du = dx$ and $v = \sin(x)$. $\int x\cos(x)\,dx = x\sin(x) - \int \sin(x)\,dx = x\sin(x) + \cos(x)$. We'll add the constant of integration in the next step. So, $\int x^2 \sin(x)\,dx = -x^2 \cos(x) + 2\int x\cos(x)\,dx = -x^2 \cos(x) + 2[x\sin(x) + \cos(x)] + C = -x^2 \cos(x) + 2x\sin(x) + 2\cos(x) + C.$ In this case we used integration by parts twice.

* This is a BC-only topic.

iii. To solve $\int e^x \sin(x)dx$, let $u = e^x$ and $dv = \sin(x)dx$.

Then, $du = e^x dx$ and $v = -\cos(x)$.

So, $\int e^x \sin(x)dx = -e^x \cos(x) + \int e^x \cos(x)dx$. We must

use integration by parts once more for $\int e^x \cos(x)dx$. Let

$u = e^x$ and $dv = \cos(x)dx$. Then $du = e^x dx$ and $v = \sin(x)$.

$\int e^x \cos(x)dx = e^x \sin(x) - \int e^x \sin(x)dx$. Finally, the original

problem, $\int e^x \sin(x)dx = -e^x \cos(x) + \int e^x \cos(x)dx =$

$-e^x \cos(x) + e^x \sin(x) - \int e^x \sin(x)dx$. More

simply, $\int e^x \sin(x)dx = -e^x \cos(x) + e^x \sin(x) - \int e^x \sin(x)dx$.

Adding $\int e^x \sin(x)dx$ to both sides of the equation yields

$2\int e^x \sin(x)dx = -e^x \cos(x) + e^x \sin(x)$

$\rightarrow \int e^x \sin(x)dx = \dfrac{-e^x \cos(x) + e^x \sin(x)}{2} + C.$

C. *Integration by partial fractions** is used to separate a fraction with a factored denomination into the sum of two fractions, each with a denominator as one of the factors; For instance,

$$\int \frac{1}{(x-1)(x+1)}dx = \int\left(\frac{A}{x-1} + \frac{B}{x+1}\right)dx$$

Find A: Ignoring $(x-1)$, let $x = 1$ in $\dfrac{1}{(x-1)(x+1)}$ getting $A = \dfrac{1}{2}$

Find B: Ignoring $(x+1)$, let $x = -1$ in $\dfrac{1}{(x-1)(x+1)}$ getting $A = \dfrac{-1}{2}$

$$\int \frac{1}{(x-1)(x+1)}dx = \int\left(\frac{\frac{1}{2}}{(x-1)} + \frac{-\frac{1}{2}}{(x+1)}\right)dx =$$

$\dfrac{1}{2}\ln|x-1| - \dfrac{1}{2}\ln|x+1| = \dfrac{1}{2}\ln\left|\dfrac{x-1}{x+1}\right| + C$

Use this method when the integrand is a fraction with non-repeating linear factors in the denominator and *u*-substitution cannot be used.

* This is a BC-only topic.

Keep in Mind...

➤ Remember that $\int \dfrac{1}{1-x} dx \neq \ln|1-x| + C$ but

$\int \dfrac{1}{1-x} dx = -\ln|1-x| + C$. This is often part of the method of partial fractions.

➤ When doing integration by parts, let dv be equal to the factor which is simpler to integrate. If you are not sure which factor to let equal u and which to let equal dv, and the integration becomes more cumbersome instead of simpler, then you've picked the factors incorrectly. Switch them.

➤ There is no equivalent integration rule for the derivative quotient rule. But an integral expression with a polynomial numerator and monomial denominator can be split into separate integrals. For example $\int \dfrac{x^2 + 4x - 1}{x} dx = \int \left(x + 4 - \dfrac{1}{x} \right) dx$.

➤ When deciding which integration method to use, eliminate the possibilities in order from easiest to most difficult—working backwards: u-substitution, integration by parts (generally used for a product of functions), integration by partial fractions (generally used for rational functions in which the denominator is a linear or quadratic function with non-repeating factors).

CHAPTER 7
PRACTICE PROBLEMS
(See solutions on page 228)

1. Find the area bounded by $y = 1 - x^2$ and the x-axis on [0, 2].

2. Given $\int_a^b f(x)dx = 5$ and $\int_a^b g(x)dx = -3$ evaluate:

(A) $\int_a^b [3f(x) - 2g(x)]dx$

(B) $\int_b^a 6f(x)dx + \int_a^a \frac{g(x)}{\pi}dx$

3. (A) Find $\dfrac{d}{dx}\int_0^x \sqrt{3t+1}\,dt$

(B) Find $\dfrac{d}{dx}\int_0^{4x^2} \dfrac{1}{e^t}\,dt$

4. Evaluate: $\int_4^\infty e^{-x}\,dx$

5. Approximate the area between $y = \sqrt{x+4}$ and the x-axis from $x = -3$ to $x = 2$ using 5 equal subdivisions by using

(A) LRAM
(B) RRAM
(C) MRAM
(D) TRAP

Integrate and state the method used:

6. $\int e^x \sqrt{e^x}\,dx$

7. $\int \sin^3(4x)\cos(4x)\,dx$

8. $\int x \ln x\,dx$

9. $\int \dfrac{2x}{x^2 - 3x + 2}\,dx$

10. $\int \dfrac{2}{3 + x^2}\,dx$

Differential Equations

I. DIFFERENTIAL EQUATIONS AND SLOPE FIELDS

A. *Differential equations* are equations that contain at least one derivative. To solve a differential equation means to find the original function. That is, given an equation containing $f'(x)$, you must find $f(x)$. This implies at some point taking the antiderivative of $f'(x)$.

1. To solve a differential equation, separate and integrate. That is, algebraically manipulate the equation such that all the x terms are on one side of the equation and all the y terms are on the other. Then integrate each side. Each side will yield a constant of integration but combining them yields only one.

   i. For example, solve $\dfrac{dy}{dx} = \dfrac{1}{y}$ given that $y\,|_{x=1} = 3$. Separating the variables yields $y\,dy = dx$. Taking the antiderivative of both sides, $\int y\,dy = \int dx \rightarrow \dfrac{y^2}{2} + C_1 = x + C_2$. The two constants of integration get combined to yield $\dfrac{y^2}{2} = x + C_2 - C_1$ or simply, $\dfrac{y^2}{2} = x + C$. There is no need to write the two constants; writing only one C is acceptable. To find the value of C apply the initial condition, $y\,|_{x=1} = 3$. This yields: $\dfrac{3^2}{2} = 1 + C \rightarrow C = \dfrac{7}{2} \rightarrow \dfrac{y^2}{2} = x + \dfrac{7}{2}$. The equation may be left as is or it may be algebraically manipulated into one of the various other forms: $y^2 = 2x + 7$, or $y^2 - 2x = 7$, $y = \sqrt{2x + 7}$, but not $y = \pm\sqrt{2x + 7}$ as the point (1, 3) doesn't pass through $y = -\sqrt{2x + 7}$.

B. Slope fields are fields of slopes, literally. Given a family of differentiable functions, $y = f(x) + C$, imagine drawing a tiny tangent line at each point on these functions. The set of all of these tangent lines forms the slope field for the function. When given a differential equation, the original function can be obtained by drawing the slope field.

1. For example, given $\dfrac{dy}{dx} = -\dfrac{x}{y}$ and a point on the original function, $(\sqrt{2}, \sqrt{2})$. Substituting some x and y values into $\dfrac{dy}{dx} = -\dfrac{x}{y}$ helps us create the slope field, below.

x	y	$m = \dfrac{dy}{dx} = -\dfrac{x}{y}$
0	1	0
0	2	0
0	−1	0
0	−2	0
1	0	$-\infty$
1	1	−1
1	2	$-\dfrac{1}{2}$
1	−1	1
1	−2	$\dfrac{1}{2}$
−1	0	∞
−1	1	1
−1	2	$\dfrac{1}{2}$
−1	−1	−1
−1	−2	$-\dfrac{1}{2}$

x	y	$m = \dfrac{dy}{dx} = -\dfrac{x}{y}$
2	0	$-\infty$
2	1	-2
2	-1	2
-2	0	∞
-2	1	2
-2	-1	-2

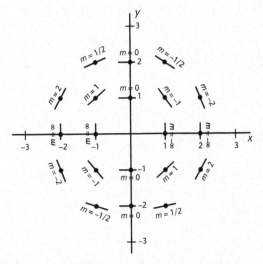

This slope field suggests that the function whose derivative is given belongs to the family of circles with the center at the origin.

Solving the differential equation by separating and integrating, yields

$$\frac{dy}{dx} = -\frac{x}{y} \rightarrow y\,dy = -x\,dx \rightarrow \int y\,dy = -\int x\,dx \rightarrow \frac{y^2}{2} = -\frac{x}{2} +$$

$$C \xrightarrow{\text{substitute in } (\sqrt{2},\sqrt{2}) \text{ to find } C} \frac{\left(\sqrt{2}\right)^2}{2} = -\frac{\left(\sqrt{2}\right)^2}{2} + C \rightarrow C = 2.$$

The particular solution to the given differential equation is $\dfrac{y^2}{2} = -\dfrac{x^2}{2} + 2$, or equivalently, $x^2 + y^2 = 4$, which represents a circle with the center at the origin and radius 2 units.

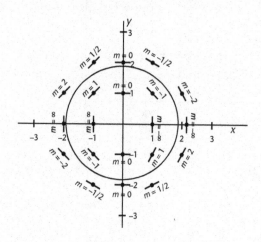

II. EULER'S METHOD*

A. This method is used for approximating values of a function given a point on the function, the function's derivative, and the step size for x (the smaller the step size, the better the approximations).

B. Using the given point on a function, (x_0, y_0), the function's derivative, and the step size for x, Δx, one can approximate the y values of the function at $x_1 = x_0 + \Delta x$, $x_2 = x_1 + 2\Delta x$, $x_3 = x_2 + 3\Delta x$ and so on. Once again, starting with the equation of the tangent line to $f(x)$ at (x_0, y_0), we have: $y_1 - y_0 = m(x_1 - x_0) \rightarrow y_1 - y_0 = m\Delta x$ $\rightarrow y_1 = y_0 + m\Delta x \rightarrow f(x_1) = f(x_0) + f'(x_0)\Delta x$. In general, $f(x_{n+1}) = f(x_n) + f'(x_n)\Delta x$.

Test Tip *If you thoroughly understand the concept and do not wish to memorize yet another formula, you can always use the equation of the tangent line to approximate the values of y.*

* This is a BC-only topic.

Example: Let $f(x) = x^2$. Use Euler's method to approximate $f(2.6)$ given that $f(2) = 4$, $\dfrac{dy}{dx} = 2x$ and $\Delta x = 0.2$. We need to calculate $f(2.2)$ which will help us calculate $f(2.4)$ which will help us calculate $f(2.6)$. Here, $x_0 = 2$. So, $f(2.2) = f(2) + f'(2)\Delta x \rightarrow f(2.2) = 4 + 2(2)(0.2) \rightarrow f(2.2) = 4.8$. Repeating this process with $x_1 = 2.2$, we have: $f(2.4) = f(2.2) + f'(2.2)\Delta x \rightarrow f(2.4) = 4.8 + 2(2.2)(0.2)$ $\rightarrow f(2.4) = 5.68$. One last iteration, with $x_2 = 2.4$: $f(2.6) = f(2.4) + f'(2.4)\Delta x \rightarrow f(2.6) = 5.68 + 2(2.4)(0.2)$ $\rightarrow f(2.6) = 6.64$. The approximation becomes increasingly less accurate as x gets larger because the error gets larger with every iteration. Below is a graphical representation of the original function, $y = x^2$ and its approximation:

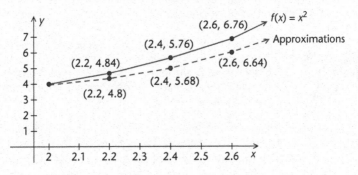

Keep in Mind...

➤ Do not round off your answers until the last step of the problem. And round off to 3 decimal places.

➤ Do only the number of iterations asked in the problem.

III. EXPONENTIAL GROWTH AND DECAY AND LOGISTIC GROWTH

A. *Exponential growth and decay* refer to the change in a quantity over time.

   1. The relationship between this quantity (bacteria, population, money in a bank account, etc.) and time is represented by the differential equation $\frac{dy}{dt} = ky$ (this states that the change in the quantity over time is proportional to the amount present at any time t whose solution is $y = y_0 e^{kt}$ where y represents the quantity at time t, y_0 represents the initial quantity, and k is a constant. If $k > 0$ then these equations represent exponential growth; if $k < 0$ these equations represent exponential decay.

 Test Tip

> *The words and phrases "quantity increases exponentially," "quantity decreases exponentially," "change in quantity is proportional to the amount present," "exponential growth," "exponential decay," or simply "bacteria growth" indicate that you must use $y = y_0 e^{kt}$.*

     i. For example, originally there are 10 bacteria in a dish. Four hours later there are 15 bacteria in the dish. How long will it take for the number of bacteria to reach 30? Since the problem involves bacteria, it must fall into the category of exponential growth. Here, $y = 15$, $y_0 = 10$, $t = ?$ when $y = 30$. Hence, $y = y_0 e^{kt} \rightarrow 15 = 10e^{k(4)} \rightarrow 1.5 = e^{4k} \rightarrow \ln(1.5) = 4k \rightarrow$

$k = \dfrac{\ln(1.5)}{4}$. Then, $30 = 10e^{\frac{\ln(1.5)}{4}t} \rightarrow 3 = e^{\ln(1.5)\frac{t}{4}} \rightarrow$

$3 = 1.5^{\frac{t}{4}} \rightarrow \ln(3) = \dfrac{t}{4}\ln(1.5) \rightarrow t = 10.838$ hours.

B. *Logistic growth** occurs when there are limiting factors present.

   1. For instance, the population of fish in a fish tank is growing logistically (as opposed to exponentially which implies no bound) because there are factors that slow the growth down

* This is a BC-only topic.

such as the competition for space, food, and oxygen. At some point, the population will level off.

To recognize questions involving logistic growth, look for the terms "logistic growth" or spot the equations involved:

$$\frac{dP}{dt} = \frac{k}{M}P(M - P) \text{ or its solution, } P(t) = \frac{M}{1 + Ae^{-kt}}, \text{ where } P \text{ is}$$

the size of the population at any time t, k and A are constants, and M is the maximum population (also called the carrying capacity).

Some facts about logistic growth:

(a) The population grows fastest at $P = \dfrac{M}{2}$ (this is where $P(t)$ has an inflection point which is the maximum of $\dfrac{dP}{dt}$)

(b) $\lim\limits_{t \to \infty} P(t) = M$ ($y = M$ is also the horizontal asymptote of $P(t)$)

(c) The equation of the population is $P(t) = \dfrac{M}{1 + Ae^{-kt}}$ and its graph has an S-shape as shown below.

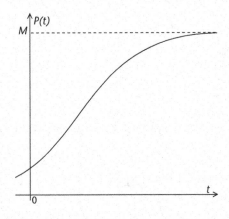

Keep in Mind...

➤ Memorize the differential equation and its solution for exponential growth/decay so that you do not have to derive the solution.

➤ While you are not responsible for the solution to the differential equation signaling logistic growth, you are responsible for the differential form of the equation as well as its component parts. You must also know the shape of a logistic growth curve as well at the location where the growth is occurring at the fastest rate.

CHAPTER 8
PRACTICE PROBLEMS
(See solutions on page 233)

For problems 1 and 2, find the general solution to the given differential equation.

1. $\dfrac{dy}{dx} = \dfrac{x}{y}$

2. $\dfrac{dy}{dx} = \dfrac{y}{x}$

3. Create a slope field for x and y taking on integers from -2 to 2 for the DEQ $\dfrac{dy}{dx} = x - y$, with 2 specific solutions passing through the points $(0, 0)$ and $(0, -1)$

4. Gargi, a bristlecone pine tree in the White Mountains of California, is the oldest known tree at age 5,065 years. If the half-life of carbon 14 is 5,750 years (100 units of carbon 14 will decay to 50 in 5,750 years), determine what percentage of carbon 14 you would expect to find in the innermost ring of Gargi.

5. Solve: $\dfrac{dy}{dx} = 2x(y+1)$ given that $y(0) = 1$.

6. The growth of a given population is represented by
$\dfrac{dP}{dt} = \dfrac{3}{10}P(10-P)$ where P represents the population in millions.

(A) When does the population grow fastest?

(B) Evaluate: $\lim\limits_{t \to \infty} P(t)$ and explain the meaning of this answer.

Applications of Integration

I. AVERAGE VALUE

A. On the first graph of $f(x)$ below, we know that the area under the curve between $x = 0$ and $x = 2$ is $\int_{0}^{2} f(x)$. Question: Is there a rectangle with base 2 whose area is exactly equal to the value of this definite integral?

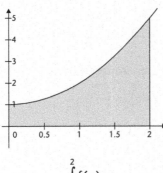

$$\int_{0}^{2} f(x)$$

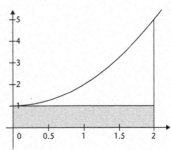

Clearly, this rectangle's area is less than $\int_{0}^{2} f(x)$.

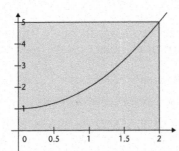

Clearly, this rectangle's area is greater than $\int_{0}^{2} f(x)$.

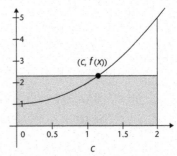

The area of this rectangle is close to $\int_{0}^{2} f(x)$.

B. This is summarized by the **Mean Value Theorem for Integrals** (not the same as the MVT for derivatives).

If *f* is continuous on the closed interval [*a*, *b*], there exists a number *c* on [*a*, *b*] such that $\int_a^b f(x)dx = f(c)(b-a)$. What this says is that the area under the curve $f(x)$ between $x = a$ and $x = b$ can be expressed as the area of a rectangle with base $(b - a)$ times the height of the rectangle at some point $(c, f(c))$.

Since $\int_a^b f(x)dx = f(c)(b-a)$, it follows that $f(c) = \dfrac{\int_a^b f(x)dx}{b-a}$, which is called the **average value** of the function. Geometrically, it represents the height of the rectangle at the point $(c, f(c))$.

Example: Given the function $f(x) = x^2 + 1$ on [0, 2], find

a) the average value of $f(x)$

$$f_{avg} = \frac{\int_0^2 (x^2 + 1)\, dx}{2-0} = \frac{\left[\frac{x^3}{3} + x\right]_0^2}{2} = \frac{\frac{8}{3} + 2}{2} = \frac{7}{3}$$

b) the value of *c* guaranteed by the MVT for integrals

$$x^2 + 1 = \frac{7}{3} \Rightarrow x = \frac{4}{3} \Rightarrow x = \pm\frac{2}{\sqrt{3}}$$

Only $\dfrac{2}{\sqrt{3}}$ is in [0, 2]

II. MOTION REVISITED

A. *Rectilinear motion in the Cartesian system.* Since the acceleration, velocity, and displacement of an object can be expressed as derivatives, that is, $v(t) = \dfrac{ds}{dt}$ and $a(t) = \dfrac{dv}{dt} = \dfrac{d^2s}{dt^2}$, it follows that they may also be expressed as antiderivatives.

1. The total *displacement* of an object moving in a straight line from $t = t_1$ to $t = t_2$ is represented by $\int_{t_1}^{t_2} v(t)dt$.

The displacement equation is given by $\int v(t)dt$ and requires an initial condition to be given. Loosely speaking, the displacement equals the integral of velocity.

i. For example, an object travels with velocity given by $v(t) = 2t^2 - 2t$ and it is given that $s(0) = 1$. Its displacement equation is given by $s(t) = \int (2t^2 - 2t)\,dt = \dfrac{2t^3}{3} - t^2 + C$. Since

$s(0) = 1,\ 1 = \dfrac{2(0)^3}{3} - 0^2 + C\ \rightarrow\ 1 = C\ \rightarrow\ s(t) = \dfrac{2t^3}{3} - t^2 + 1.$

ii. The total displacement of the object in part i for the first two seconds is given by $\displaystyle\int_0^2 (2t^2 - 2t)\,dt = \left.\dfrac{2t^3}{3} - t^2\right]_0^2 = 1.\bar{3}$

2. The total distance traveled by an object on the interval (t_1, t_2) is represented by $\displaystyle\int_{t_1}^{t_2} |v(t)|\,dt$. Loosely speaking, the total distance equals the integral of the speed.

i. For example, the total distance traveled by the object in part i, above, is given by $d = \displaystyle\int_0^2 |(2t^2 - 2t)|\,dt = 2$.

III. INTEGRATION APPLICATIONS

A. Derivatives have applications of motion, related rates, and optimization. Integration also have applications that incorporate the rules below.

Integral of the rate of change to give accumulated change:

$$\int_a^b R'(t)\,dt = \left[R(t)\right]_a^b = R(b) - R(a) \text{ or } R(b) = R(a) + \int_a^b R'(t)\,dt$$

Derivative of the accumulation function: $\dfrac{d}{dt}\left[\displaystyle\int_a^x f(t)\,dt\right] = f(x)$

1. *Example:* It is early May and Philadelphia experiences a brutal heat wave that lasts 4 days. Many people turn on their air conditioners and find that they don't work. They start calling for service and the service companies can't handle the volume of requests and must schedule appointments many days later. One of those companies charts $r(t)$, the

rate that service requests come into the company, measured in requests per day, as shown in the figure below. Time t is measured in days starting when the heat wave starts. The company can do 20 service calls a day and already had 10 people waiting for service at $t = 0$.

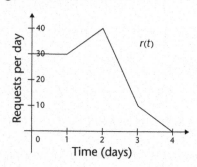

i. How many people have requested service by the end of the heat wave?

$$10 + \int_0^4 r(t)dt = 10 + 30 + \frac{1}{2}(30 + 40) + \frac{1}{2}(40 + 10) + \frac{1}{2}(10) = 105$$

ii. How many people are waiting for service at the end of the heat wave?
Requests = 105
Service calls = 20(4) = 80
25 people are still waiting

iii. Is the number of people waiting for service increasing, decreasing, or staying the same on the first day of the heat wave? Explain.
It is increasing because requests are being processed at 20 a day for $0 < t < 1$, $r(t) > 20$.

iv. On what day was the number of people waiting for service the longest? How many people are waiting at that time? Justify your answer.

$$W(t) = 10 + \int_0^t r(x)\, dx - 20t \text{ so } W'(t) = r(t) - 20 = 0$$

$r(t) > 20$ for $t < 2\frac{2}{3}$, $r(t) < 20$ for $t > 2\frac{2}{3}$

So the number of people waiting for service is greatest $t = 2\frac{2}{3}$, day 3 of the heat wave. There will be

$$10 + \int_0^{2\frac{2}{3}} r(x)\, dt - 20(2\frac{2}{3}) = 10 + 30 + 35 + \frac{1}{3}(40 + 20) - 53\frac{1}{3}$$

$$\approx 42 \text{ people.}$$

IV. AREA IN CARTESIAN COORDINATES

A. Area between a curve and the *x*-axis

1. The area between $f(x)$ and the *x*-axis, if $f(x) \geq 0$ from $x = a$ to $x = b$, is represented by $\int_a^b f(x)dx$.

   i. For instance, the area between $f(x) = \sqrt{x}$ and the *x*-axis on [0, 4], is given by $\int_0^4 \sqrt{x}\,dx = 5.\bar{3}$.

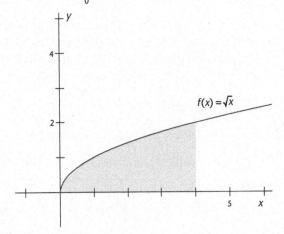

2. The area between $f(x)$ and the *x*-axis, if $f(x) \leq 0$ from $x = a$ to $x = b$, is represented by $\left| \int_a^b f(x)dx \right|$.

   i. For instance, the area between $f(x) = x^2 - 4$ and the *x*-axis on [−2, 2], is given by $\left| \int_{-2}^2 (x^2 - 4)\,dx \right| = 10.\bar{6}$.

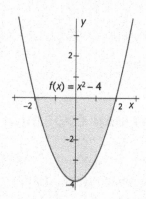

$f(x) = x^2 - 4$

3. The area between $f(x)$ and the x-axis, if $f(x)$ is sometimes negative and sometimes positive from $x = a$ to $x = b$, is represented by $\int_a^b |f(x)|\,dx$. Evaluate the integral on the interval(s) on which $f(x) \geq 0$, and add it to the absolute value of the integral on the intervals(s) on which $f(x) \leq 0$. That is, suppose that $f(x) \geq 0$ on $[a, b]$ and $f(x) < 0$ on $[c, d]$ where $a < b < c < d$. The area between $f(x)$ and the x-axis on $[a, d]$ is given by $\int_a^b f(x)\,dx + \left| \int_c^d f(x)\,dx \right|$.

    i. For instance, the area between $f(x) = 1 - x^2$ and the x-axis on $[0, 3]$, is given by either $\int_0^3 |1 - x^2|\,dx = 7.\overline{3}$ or,

    equivalently, $\int_0^1 (1 - x^2)\,dx + \left| \int_1^3 (1 - x^2)\,dx \right| = 7.\overline{3}$.

4. To find the area between $f(x)$ and the y-axis on $y = c$ to $y = d$, rewrite the equation in terms of y first. That is, rewrite the equation in the form $x = g(y)$. If $g(y) > 0$ the area is represented by $\int_c^d g(y)\,dy$. If $g(y) < 0$ the area is represented by $\left| \int_c^d g(y)\,dy \right|$. If $g(y) > 0$ on $[a, b]$ and $g(y) < 0$ on $[c, d]$

the area between $g(y)$ and the y-axis on $[a, d]$ is given by

$$\int_a^b g(y)dy + \left|\int_c^d g(y)dy\right| \text{ or, equivalently, } \int_a^d |g(y)|dy.$$

i. For instance, the area between $f(x) = e^x$ and the y-axis from $y = 1$ to $y = 2$, is given by $\int_1^2 (\ln(y))dy = .386$. Note that $f(x) = e^x \rightarrow y = e^x \rightarrow x = \ln(y)$.

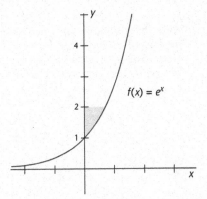

ii. Similarly, the area between $f(x) = e^x$ and the y-axis from $y = \dfrac{1}{2}$ to $y = 1$, is given by $\left|\int_{\frac{1}{2}}^1 (\ln(y))dy\right| = .153$.

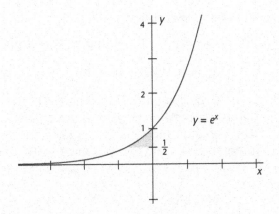

iii. Also, the area between $f(x) = e^x$ and the y-axis from $y = \dfrac{1}{2}$ to $y = 2$, is given by

$$\left|\int_{\frac{1}{2}}^{1}(\ln(y))\,dy\right| + \int_{1}^{2}(\ln(y))\,dy = .540 \text{ or}$$

$$\int_{\frac{1}{2}}^{2}|\ln(y)|\,dy = .540.$$

Test Tip

Note that although these two methods are equivalent mathematically, the calculator is using a variation of Riemann sum calculations and thus an approximation, but it will always be accurate to 3 decimal places.

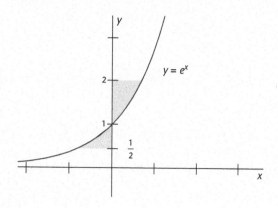

B. Area between two curves

1. The area between $f(x)$ and $g(x)$, where $g(x) \le f(x)$ from $x = a$ to $x = b$, is represented by $\displaystyle\int_{a}^{b}(f(x) - g(x))\,dx$.

   Loosely speaking, this is the integral of the top function minus the bottom function. If the answer is negative, then this is an indication that the order of the functions in the integrand is wrong and you must switch the functions around.

   i. For instance, the area between $f(x) = x$ and $g(x) = x^2$ on $[2, 3]$ is given by $\displaystyle\int_{2}^{3}\left(x^2 - x\right)dx = 3.8\overline{3}$.

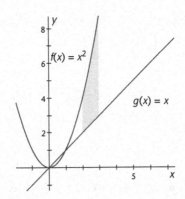

2. The area between $f(y)$ and $g(y)$, where $g(y) \leq f(y)$ from $y = c$ to $y = d$, is represented by $\int_{c}^{d}(f(y) - g(y))dy$. Loosely speaking, this is the integral of the right function minus the left function. If the answer is negative, then this is an indication that the order of the functions in the integrand is wrong and you must switch the functions around.

   i. For instance, the area between $f(y) = y$ and $g(y) = e^y$ on $1 \leq y \leq 2$, is given by $\int_{1}^{2}(e^y - y)dy = 3.17$.

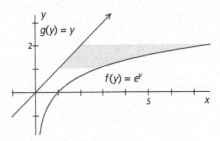

V. LENGTH OF CURVE*

A. Length of curve in Cartesian coordinates

    1. The length (also known as arc length) of a smooth

* This is a BC-only topic.

Cartesian curve, $f(x)$, from $x = x_1$ to $x = x_2$, is represented

by $L = \int_{x_1}^{x_2} \sqrt{1 + \left(\dfrac{dy}{dx}\right)^2}\, dx$

i. For example, the length of $f(x) = x^2$ on $[0, 1]$ is given by

$L = \int_0^1 \sqrt{1 + (2x)^2}\, dx = 1.479$.

2. The length of a smooth Cartesian curve, $f(y)$, from $y = y_1$ to

$y = y_2$ is represented by $L = \int_{y_1}^{y_2} \sqrt{1 + \left(\dfrac{dx}{dy}\right)^2}\, dy$

i. For example, the length of $f(y) = e^y + y$ from $y_1 = 2$ to

$y_2 = 3$ is given by $L = \int_2^3 \sqrt{1 + (e^y + 1)^2}\, dy = 13.736$.

VI. VOLUME

A. *Washer Method*—The washer method is used when the cross sections of the solid are washers (generally when the volume is that of a solid which has been created by rotating a region bounded by <u>two</u> curves about an axis).

1. If rotating the region between $f(x)$ and $g(x)$, where $f(x) \geq g(x)$, <u>about the x-axis</u>, the volume of the resulting solid is

represented by $\pi \int_{x_1}^{x_2} [(f(x))^2 - (g(x))^2]\, dx$ where x_1 and x_2

represent the x-values of the intersection points of the two functions. Loosely speaking, this is the integral of the top function squared minus the bottom function squared. Don't forget the π!

i. For example, the volume of the solid formed when the region between $f(x) = 3$ and $g(x) = x^2$ is revolved about the

x-axis, is given by $\pi \int_{-\sqrt{3}}^{\sqrt{3}} [(3)^2 - (x^2)^2]\, dx = 78.356$.

To find the x-values of the intersection points, set the

functions equal to each other and solve for x. In this case, $3 = x^2 \rightarrow x = \pm\sqrt{3}$.

2. If rotating the region between $f(x)$ and $g(x)$, where $f(x) \geq g(x)$, about a <u>horizontal line, $y = k$</u>, the volume of the resulting solid is represented by $\pi\int\limits_{x_1}^{x_2}[(f(x)-k)^2 - (g(x)-k)^2]dx$

   Don't forget the π!

   i. For example, the volume of the solid formed when the region between $y = x^2 + 4$ and $y = x + 4$ is revolved about the line $y = 1$, is given by

   $$\pi\int\limits_{0}^{1}[(x+4-1)^2 - (x^2+4-1)^2]dx = 3.560.$$

3. If rotating the region between $f(y)$ and $g(y)$, where $f(y) \geq g(y)$, <u>about the y-axis</u>, the volume of the resulting solid is represented by $\pi\int\limits_{y_1}^{y_2}[(f(y))^2 - (g(y))^2]dy$. Loosely speaking, this is the integral of the right function squared minus the left function squared. Don't forget the π!

   i. For instance, the volume of the solid formed by revolving the region bounded by $x = 3$ and $x = \ln(y)$ from $y = 1$ to $y = 3$ about the y-axis, is given by

   $$\pi\int\limits_{1}^{3}[(3)^2 - (\ln(y))^2]dy = 53.315.$$

4. When rotating the region between $f(y)$ and $g(y)$, where $f(y) \geq g(y)$, about a <u>vertical line, $x = k$</u>, the volume of the resulting solid is represented by $\pi\int\limits_{y_1}^{y_2}[(f(y)-k)^2 - (g(y)-k)^2]dy$

   or, equivalently, $\pi\int\limits_{y_1}^{y_2}[(k-f(y))^2 - (k-g(y))^2]dy$, where y_1 and y_2 represent the y-values of the intersection points of the two functions. Don't forget the π!

   i. For example, the volume of the solid obtained by rotating $x = 3$ and $x = \ln(y)$ on $1 \leq y \leq 2$ about the line $x = 4$, is given by $\pi\int\limits_{1}^{2}[(4-\ln(y))^2 - (4-3)^2]dy = 38.007.$

B. *Disk Method*—The disk method is used when the cross sections of the solid are disks. This method generally involves only one function. This is a simple case of the washer method in which $g(x) = 0$ or $g(y) = 0$. For instance, the volume of the solid formed by revolving $y = x^3$ about the x-axis on [0, 2] is given by

$$\pi\int_0^2 [(x^3)^2 - (0)^2]dx = \pi\int_0^2 x^6 dx = 57.446.$$

CHAPTER 9
PRACTICE PROBLEMS

(See solutions on page 235)

1. Find the average value of the function $f(x) = \sqrt{x}$ on the interval [1, 9].

2. Find the value of c guaranteed by the Mean Value Theorem for Integrals for $f(x) = 4 - x^2$ on [0, 2].

3. Given the velocity of a particle in ft/sec, find the displacement and distance traveled in the given time interval.

   $v(t) = 12 - 3t$ [0, 5]

4. Find the area of the region bounded by the graphs of $y = 6 - x^2$ and $y = x$. Find the points of intersection, sketch the curves, set up the integral and solve.

5. Key West Shave Ice is a popular location for a cold treat on a hot day on the beach. People have been known to line up for an hour to get served. One day, the store opened at 11 AM and there were already 12 people in line. The rate that people enter the line to be served is given by the function $E(t)$, measured in people per minute, where t is measured in minutes, $0 \le t \le 45$. The graph of $E(t)$ is shown in the figure below. How many people have waited in line at the end of 45 minutes? Figure to the nearest person.

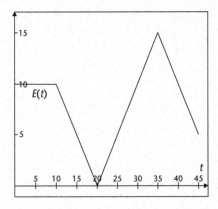

6. Find the length of the curve represented by $f(x) = e^x + 1$ on $[0, 1]$.

7. Write an expression that represents the volume when the area to the right of the y-axis and between the graphs of $y = \cos x$, $x = 0$, and $y = -1$ is rotated about $y = -1$.

PART V
CALCULUS BC TOPICS

Parametric Equations, Polar Coordinates, and Vector-Valued Functions

I. PARAMETRIC AND VECTOR-VALUED FUNCTIONS

A. Parametric and vector equations are used to describe the motion of a body. They have different notations but describe the same concept. An additional variable is involved, called the parameter, usually denoted by t (for time). The parameter does not appear on the graph; it only represents the time at which a given particle is at a given point. Both parametric and vector equations are represented on the Cartesian coordinate system.

1. Parametric equations create only one graph though they contain two equations. They are denoted by: $\begin{cases} x(t) = f(t) \\ y(t) = g(t) \end{cases}$.

2. A vector-valued function is denoted by: $\langle x(t), y(t) \rangle = \langle f(t), g(t) \rangle$ or, $r(t) = (f(t))i + (g(t))j$. The difference between parametric functions and vector-valued functions is that parametric functions describe a path in the x-y plane while a vector-valued function represents a force in the direction of that path.

3. A parametric equation can be written in Cartesian form (x–y form) by using algebraic manipulation.

   i. For example, parametric equations: $\begin{cases} x(t) = t \\ y(t) = t^2 \end{cases}$, $t \geq 0$, can be written in vector form as $\langle x(t), y(t) \rangle = \langle t, t^2 \rangle$ or $r(t) = (t)i + (t^2)j$ and in Cartesian form as $y = x^2$. See graph at right.

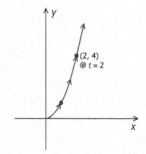

ii. The difference between a parametric/vector curve and a Cartesian curve is that a parametric/vector curve has *direction*. When drawing a parametric/vector curve, in part 3i, the direction must be specified with arrows, on the graph, in the direction of increasing parameter, or else the graph will be considered incomplete.

II. POLAR EQUATIONS

A. A polar equation, $r = f(\theta)$, is written using polar coordinates (r, θ), where r represents the point's distance from the origin, and θ represents the measurement of the angle between the positive x-axis and the line segment between the point and the origin. Angle θ is measured counterclockwise from the positive x-axis.

B. To switch from Cartesian form to polar form use: $r^2 = x^2 + y^2$ and $\theta = \tan^{-1}\left(\dfrac{y}{x}\right)$; to switch from polar form to Cartesian form, use $x = r \cos(\theta)$ and $y = r \sin(\theta)$.

   i. To change $x^2 + y^2 = 9$ from Cartesian form to polar form, rewrite $x^2 + y^2 = 9$ as $r^2 = 9 \rightarrow r = 3$ (or $r = -3$).

   ii. To change $r = 4 \sec(\theta)$ from polar form to Cartesian form, rewrite $r = 4 \sec(\theta)$ as $r = \dfrac{4}{\cos(\theta)} \rightarrow r \cos(\theta) = 4 \rightarrow x = 4$.

C. The most common polar equations with which you must be familiar are:

1. Line

• $\theta = a$ (y–axis if $a = \pm\dfrac{\pi}{2}$, x-axis if $a = 0$ or $a = \pm\pi$)
   • Vertical: $r = a \sec(\theta)$
   • Horizontal: $r = a \csc(\theta)$

Examples:

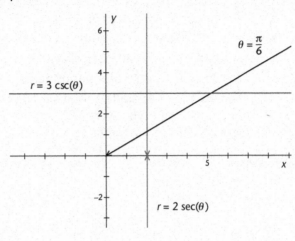

$r = 3 \csc(\theta)$

$\theta = \dfrac{\pi}{6}$

$r = 2 \sec(\theta)$

2. Circle

- With center at origin: $r = a$, length of radius is a
- Tangent to the *y*-axis, intersecting the *x*-axis: $r = a \cos(\theta)$
- Tangent to the *x*-axis, intersecting the *y*-axis: $r = a \sin(\theta)$

Examples:

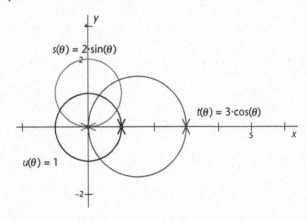

$s(\theta) = 2 \cdot \sin(\theta)$

$t(\theta) = 3 \cdot \cos(\theta)$

$u(\theta) = 1$

3. Rose

- $r = a \sin(b\theta)$ or $r = a \cos(b\theta)$
- a = length of petal from origin to opposite point
- If b is odd, then b = number of petals
- If b is even, then $2b$ = number of petals
- All petals are equidistant from each other—if there are four petals they occur every 90°, if there are three petals they occur every 120°, etc.

Examples:
$b = 2$, even, 4 petals

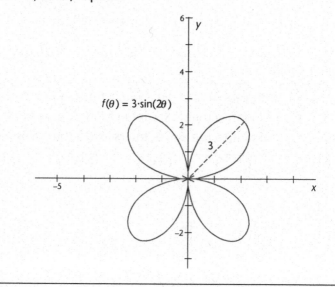

$f(\theta) = 3 \cdot \sin(2\theta)$

$b = 3$, odd, 3 petals

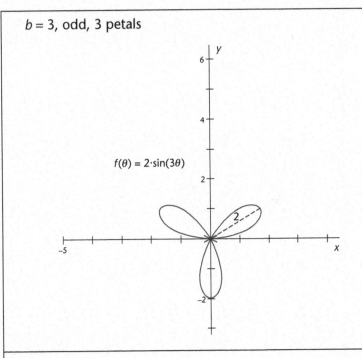

$f(\theta) = 2 \cdot \sin(3\theta)$

4. Limaçon

$r = a \pm b \sin(\theta)$ or $r = a \pm b \cos(\theta)$

- The distance from origin to farthest point from origin is $|a| + |b|$.
- If $|a| > |b|$ the limaçon is dimpled (dimple's distance from the origin is $||a| - |b||$).
- If $|a| < |b|$ the limaçon is looped (length of loop is $||a| - |b||$).
- If limaçon equation contains 'sin' and '+', the graph lies mostly above the x-axis.
- If limaçon equation contains 'sin' and '–', the graph lies mostly below the x-axis.
- If limaçon equation contains 'cos' and '+', the graph lies mostly to the right of the y-axis.
- If limaçon equation contains 'cos' and '–', the graph lies mostly to the left of the y-axis.

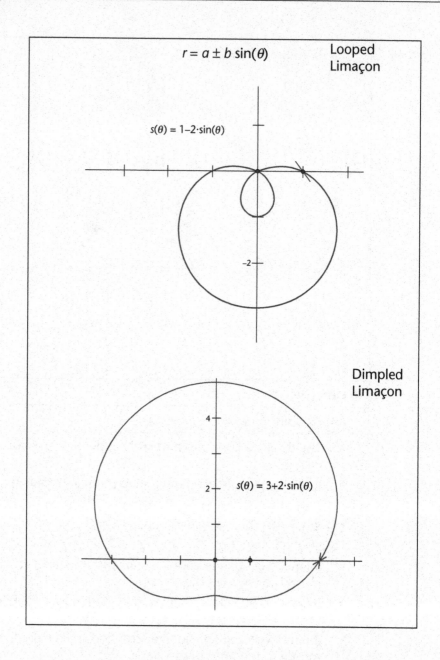

$r = a \pm b \sin(\theta)$ Looped Limaçon

$s(\theta) = 1 - 2 \cdot \sin(\theta)$

Dimpled Limaçon

$s(\theta) = 3 + 2 \cdot \sin(\theta)$

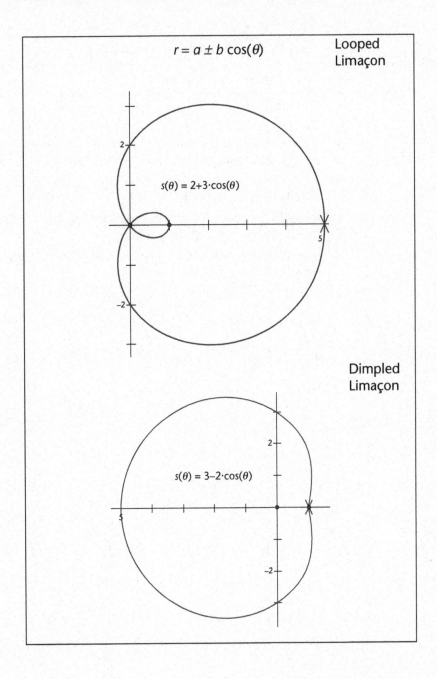

$r = a \pm b \cos(\theta)$

Looped Limaçon

$s(\theta) = 2 + 3 \cdot \cos(\theta)$

Dimpled Limaçon

$s(\theta) = 3 - 2 \cdot \cos(\theta)$

5. Cardioid (type of limaçon in which *a* = *b*)

$r = a \pm a \sin(\theta)$ or $r = a \pm a \cos(\theta)$

- The distance from origin to farthest point from origin is $2|a|$.
- If cardioid equation contains 'sin' and '+', the graph lies mostly above *x*-axis.
- If cardioid equation contains 'sin' and '−', the graph lies mostly below the *x*-axis.
- If cardioid equation contains 'cos' and '+', the graph lies mostly to the right of the *y*-axis.
- If cardioid equation contains 'cos' and '−', the graph lies mostly to the left of the *y*-axis.

$r = a \pm a \sin(\theta)$

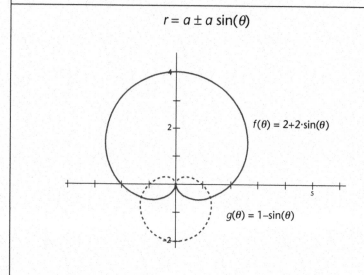

$f(\theta) = 2 + 2 \cdot \sin(\theta)$

$g(\theta) = 1 - \sin(\theta)$

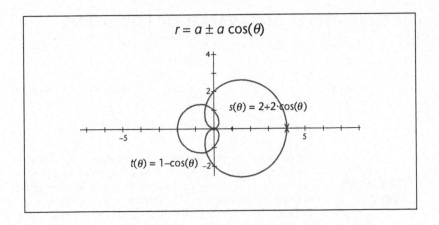

$$r = a \pm a \cos(\theta)$$

$s(\theta) = 2 + 2 \cdot \cos(\theta)$

$t(\theta) = 1 - \cos(\theta)$

Keep in Mind...

➤ Indicate the direction of motion on graphs represented by parametric equations or vector equations.

➤ When graphing parametric equations or vector-valued functions, take into account the restriction on the parameter.

➤ When using $\theta = \tan^{-1}\left(\dfrac{y}{x}\right)$ to calculate the reference angle for a point in Cartesian form, beware of pairing the right r value with the right θ value. For example, to write the Cartesian point $(-1, \sqrt{3})$ in polar form, $r^2 = x^2 + y^2 = 1^2 + (\sqrt{3})^2 = 4 \rightarrow r = \pm 2$ and $\theta = \tan^{-1}\left(\dfrac{\sqrt{3}}{-1}\right) = -\dfrac{\pi}{3}$. So, the point can be represented by $\left(2, \dfrac{2\pi}{3}\right)$ or by $\left(-2, -\dfrac{\pi}{3}\right)$, though there is an infinite number of representations of a point in polar form.

III. DERIVATIVES OF PARAMETRIC EQUATIONS

A. The first derivative—given $\begin{cases} x(t) = f(t) \\ y(t) = g(t) \end{cases} \rightarrow \dfrac{dy}{dx} = \dfrac{dy/dt}{dx/dt}$. For

example, $\begin{cases} x(t) = 2t^2 + 3t \\ y(t) = e^{2t} \end{cases} \rightarrow \dfrac{dy}{dx} = \dfrac{2e^{2t}}{4t+3}$. Note that the derivative

of a set of parametric equations is a function of t.

Test Tip

Remember that this derivative represents the slope of the graph at a given x value. But, when asked to find the slope at a given x value, substitute it back in to the original to find the t value corresponding to it. Then use that t value to substitute into the derivative and find the slope! The common mistake is to use the x value instead of the t value.

B. The second derivative, $\dfrac{d^2y}{dx^2} = \dfrac{d}{dx}\left(\dfrac{dy}{dx}\right) = \dfrac{\dfrac{d}{dt}\left(\dfrac{dy}{dx}\right)}{\dfrac{dx}{dt}}$.

In words, the numerator of this formula represents the

derivative of $\dfrac{dy}{dx}$ (in terms of t) while the denominator

represents the first derivative of $x(t)$. Since $\dfrac{dy}{dx} = \dfrac{2e^{2t}}{4t+3}$,

$$\dfrac{d^2y}{dx^2} = \dfrac{\dfrac{d}{dt}\left(\dfrac{2e^{2t}}{4t+3}\right)}{\dfrac{dx}{dt}} = \dfrac{\dfrac{(4t+3)(4e^{2t}) - 4(2e^{2t})}{(4t+3)^2}}{4t+3}$$

$$\dfrac{d^2y}{dx^2} = \dfrac{16te^{2t} + 4e^{2t}}{(4t+3)^3} = \dfrac{4e^{2t}(4t+1)}{(4t+3)^3}$$

Keep in Mind...

➤ When asked to find the derivative of parametric equations at a certain point, pay attention to whether you are given an x-value or a t-value and solve the problem accordingly.

➤ The second derivative of a parametric equation is tricky. Make sure you understand how the formula is used. Practice it until you get it right.

IV. DERIVATIVES OF POLAR EQUATIONS

A. Rewrite the polar equations in parametric form and use the parametric formulas.

1. First derivative—rewrite $r = f(\theta)$ in parametric

form: $\begin{cases} x(\theta) = r\cos(\theta) \\ y(\theta) = r\sin(\theta) \end{cases} \rightarrow \begin{cases} x(\theta) = f(\theta)\cos(\theta) \\ y(\theta) = f(\theta)\sin(\theta) \end{cases}$

$\dfrac{dy}{dx} = \dfrac{dy/d\theta}{dx/d\theta} = \dfrac{f(\theta)\cos(\theta) + \sin(\theta)f'(\theta)}{-f(\theta)\sin(\theta) + \cos(\theta)f'(\theta)}$. For example,

find the derivative (slope) of $f(\theta) = 2\sin(3\theta)$ at $\theta = \dfrac{\pi}{6}$

$\dfrac{dy}{dx} = \dfrac{dy/d\theta}{dx/d\theta} = \dfrac{f(\theta)\cos(\theta) + \sin(\theta)f'(\theta)}{-f(\theta)\sin(\theta) + \cos(\theta)f'(\theta)}$

$= \dfrac{2\sin(3\theta)\cos\theta + \sin(\theta)6\cos(3\theta)}{-2\sin(3\theta)\sin(\theta) + \cos(\theta)6\cos(3\theta)} \rightarrow \dfrac{dy}{dx}\Big|_{\theta=\frac{\pi}{6}} = -\sqrt{3}.$

Graphically, this represents the slope of the tangent line to

the graph of $f(\theta) = 2\sin(3\theta)$ at the point $\left(2, \dfrac{\pi}{6}\right)$.

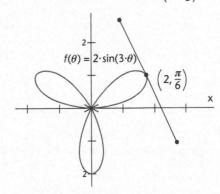

$f(\theta) = 2 \cdot \sin(3 \cdot \theta)$

$\left(2, \dfrac{\pi}{6}\right)$

V. AREA IN POLAR COORDINATES

A. Area inside a polar curve

    1. The area inside a polar curve, $r = f(\theta)$, is represented by
$$\frac{1}{2}\int_{\theta_1}^{\theta_2}(f(\theta))^2 d\theta.$$

        i. For example, the area of cardioid $r = 2 + 2\sin(\theta)$, is given
$$\text{by } \frac{1}{2}\int_0^{2\pi}(2 + 2\sin(\theta))^2 d\theta = 18.850.$$

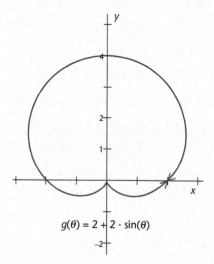

$$g(\theta) = 2 + 2 \cdot \sin(\theta)$$

B. Area between two polar curves

    1. The area between $h(\theta)$ and $g(\theta)$, if $h(\theta) > g(\theta)$ is represented
by $\frac{1}{2}\int_{\theta_1}^{\theta_2}[(h(\theta))^2 - (g(\theta))^2]d\theta$ where the limits, θ_1 and θ_2,
represent the angles at which the two curves intersect.

        i. For example, the area between $h(\theta) = 1 + \cos(\theta)$ and
$g(\theta) = 3\cos(\theta)$ in the first and fourth quadrants, is

given by $\frac{1}{2}\int_{-\frac{\pi}{3}}^{\frac{\pi}{3}}[(3\cos(\theta))^2 - (1 + 3\cos(\theta))^2]d\theta$, where

$\theta = \pm\dfrac{\pi}{3}$ are the angles at which the curves intersect.

Because of x-axis symmetry, an equivalent solution is:

$\displaystyle\int_{0}^{\frac{\pi}{3}}[(3\cos(\theta))^2 - (1 + 3\cos(\theta))^2]d\theta$, that is, double the area

from $\theta = 0$ to $\theta = \dfrac{\pi}{3}$.

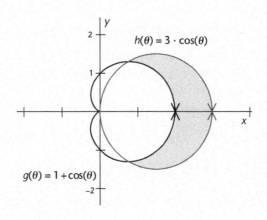

$h(\theta) = 3 \cdot \cos(\theta)$

$g(\theta) = 1 + \cos(\theta)$

VI. ARC LENGTH

A. Length of curve in parametric form

1. The length of a parametric curve, from $t = t_1$ to $t = t_2$ is

   represented by $L = \displaystyle\int_{t_1}^{t_2}\sqrt{\left(\dfrac{dx}{dt}\right)^2 + \left(\dfrac{dy}{dt}\right)^2}\, dt$

   i. For example, the length of the parametric curve

   represented by $\begin{cases} x(t) = t \\ y(t) = t^2 \end{cases}$ from $t = 1$ to $t = 3$ is given by

   $L = \displaystyle\int_{1}^{3}\sqrt{(1)^2 + (2t)^2}\, dt = 8.268.$

B. Length of curve in polar form

1. The length of a polar curve $r = f(\theta)$ between $\theta = a$ and $\theta = b$ is given by the integral $L = \int_a^b \sqrt{r^2 + \left(\dfrac{dr}{d\theta}\right)^2}\, d\theta$.

VII. MOTION ON A PARAMETRIC/VECTOR CURVE

A. In vector form, recall that the velocity of an object is given by $v = \langle x'(t), y'(t) \rangle$. Thus, the speed is given by $|v| = \sqrt{(x'(t))^2 + (y'(t))^2}\, dt$, also called the magnitude of the velocity vector. This also represents the speed in parametric form. Hence, the total distance traveled on such a curve is given by $d = \int_{t_1}^{t_2} \sqrt{(x'(t))^2 + (y'(t))^2}\, dt$ for both a parametric curve, $\begin{cases} x = x(t) \\ y = y(t) \end{cases}$, and a vector curve $\langle x(t), y(t) \rangle$.

1. For example, the total distance traveled along $\begin{cases} x = 2t \\ y = \ln(t) \end{cases}$ on $1 \le t \le 2$, is given by $d = \int_1^2 \sqrt{(2)^2 + \left(\dfrac{1}{t}\right)^2}\, dt = 2.121$.

B. Motion along a parametric/polar curve
   1. For a polar curve $r = f(\theta)$, you must rewrite the original function in parametric form and use the formulas above.

CHAPTER 10
PRACTICE PROBLEMS

(See solutions on page 236)

1. Sketch the graph given by $\begin{cases} x(t) = 2\sin(t) \\ y(t) = 3\cos(t) \end{cases}$ $\quad 0 \le t \le \pi$

2. Sketch the graph given by $r(t) = \left(\dfrac{t}{2}\right)i + (e^t)j \quad t > 0$

3. Name and sketch $r = 2 - 3\cos(\theta)$.

4. Find the number of petals and the length of each petal of $r = 4\sin(2\theta)$.

5. Write in Cartesian form: $r = \cos(\theta)$.

6. Write in polar form: $x = 2$.

7. Find $\dfrac{dy}{dx}$ and $\dfrac{d^2y}{dx^2}$ and evaluate each at the indicated value of t for $x = 2\sin t,\ y = 2\cos t \quad t = \dfrac{5\pi}{4}$.

In the following, graph on your calculator and find the area of the region using the calculator.

8. One petal of $r = 6\cos 3\theta$.

9. A particle moves along a plane curve described by $r(t) = 3\sin\left(\dfrac{t}{2}\right)i + 3\cos\left(\dfrac{t}{2}\right)j$. Find the velocity vector, acceleration vector, and speed. Interpret the speed.

Infinite Sequences and Series

I. **SEQUENCES**—a sequence is a list of numbers separated by commas $a_1, a_2, a_3, \ldots, a_k, \ldots$, that may or may not have a pattern.

A. Arithmetic and geometric sequences

1. The formula for the nth term of an arithmetic sequence (one that is formed by adding the same constant repeatedly to an initial value) is $a_n = a_1 + (n-1)d$ where a_1 is the first term of the sequence, n is the number of terms in the sequence, and d is the common difference. The formula for the nth term of a geometric sequence (one that is formed by multiplying the same constant repeatedly to an initial value) $a_n = a_1 r^{(n-1)}$ where a_1 is the first term, r is the common ratio, and n is the number of terms in the sequence.

2. Convergent sequences—a sequence converges if it approaches a number as the number of terms becomes infinite. A sequence can be thought of as a function whose domain is the set of positive integers. As such, the concept of limit of a sequence is the same as the concept of limit of a function.

For example, $\lim\limits_{n \to \infty} \left\{ \dfrac{n}{2n+3} \right\} = \dfrac{1}{2}$. For example, $\lim\limits_{x \to \infty} \left\{ (-1)^n \right\}$ DNE.

3. Divergent sequences—a sequence is divergent if it does not approach a particular number or it approaches $\pm\infty$.

For example, $\lim\limits_{n \to \infty} \left\{ \dfrac{n^2}{2n+3} \right\} = \infty$.

II. **SERIES**—A series is the sum of the terms of a sequence.

If a sequence is $a_1, a_2, a_3, \cdots, a_n$ then the series described by the sequence is $a_1 + a_2 + a_3 + \cdots + a_n = \sum\limits_{k=1}^{n} a_k$. If a sequence has a finite number of terms, the series also sums a finite number of terms and is called a partial sum. A finite sequence and finite series will always converge. However, we are mostly interested in infinite sequences and series. An infinite sequence that converges has the nth term approaching a limit. $\lim_{n \to \infty}(a_n) = L$. An infinite series that converges has the sum of its infinite terms approaching a limit. $\sum_{n=1}^{\infty} a_n = L$.

With most infinite series, it is possible only to figure out whether it converges (or diverges) but not to figure out the actual sum. In general, the series for which it is possible to find the sum, if it exists, are geometric series and telescoping series.

A. Types of infinite series

1. *Geometric series*—this series is of the form $a + ar + ar^2 + \cdots + ar^n + \ldots$ $\sum_{k=0}^{\infty} ar^k$. This series converges (that is, its sum exists) if and only if $|r| < 1$ (that is, $-1 < r < 1$). If it converges, its sum is given by $S = \dfrac{a}{1-r}$.

2. *p-series*, $\sum_{k=1}^{\infty} \dfrac{1}{k^p}$, $p > 0$, converges when $p > 1$ and diverges when $0 < p \le 1$.

3. *Alternating series* are series with terms whose signs alternate. They are of the form $\sum_{k=1}^{\infty}(-1)^k a_k$ or $\sum_{k=1}^{\infty}(-1)^{k+1}a_k$.

4. *Harmonic series*, $\sum_{k=1}^{\infty} \dfrac{1}{k}$, diverges. This is a *p*-series with $p = 1$.

5. *Alternating Harmonic series*, $\sum_{k=1}^{\infty} \dfrac{(-1)^k}{k}$, converges.

6. *Alternating p-series* $\sum_{k=1}^{\infty} \dfrac{(-1)^k}{k^p}$ converges for $p > 0$.

7. *Power series* in x: $\sum_{k=1}^{\infty} a_k x^k$. Power series in $(x - a)$: $\sum_{k=1}^{\infty} a_k(x - a)^k$. (More on power series later on.)

8. *Telescoping series* is a series in which all but a finite number of terms cancel out. It is either decomposed into partial fractions or you need to decompose it yourself. For example,

$$\sum_{k=1}^{\infty}\frac{1}{k^2+3k+2} = \sum_{k=1}^{\infty}\left(\frac{1}{k+1}-\frac{1}{k+2}\right) = \left(1-\frac{1}{2}\right)+\left(\frac{1}{2}-\frac{1}{3}\right)+$$

$$\left(\frac{1}{3}-\frac{1}{4}\right)+\cdots+\left(\frac{1}{k}-\frac{1}{k+1}\right)+\left(\frac{1}{k+1}-\frac{1}{k+2}\right) = 1-\frac{1}{k+2}. \text{ This}$$

series converges because $\lim\limits_{k\to\infty}\left(1-\frac{1}{k+2}\right) = 1.$

Not all telescoping series converge.

For example: $\sum\limits_{k=1}^{\infty} k-(k+1) = (1-2)+(2-3)+(3-4)+\cdots+$

$k-(k+1) = -1$ *and* $\lim\limits_{k\to\infty}(-1) = -1-1-1-\cdots = -\infty$, *so the*

series diverges.

B. **Convergence/Divergence Tests for Series**

Let $\sum\limits_{k=1}^{\infty}a_k$ be an infinite series of positive terms. This series

converges if for any value *n*, the series $\sum\limits_{k=1}^{n}a_k$ converges. That is

if any possible partial sum S_n converges, then the infinite series

converges. An infinite series $\sum\limits_{k=1}^{\infty}a_k$ may have negative terms. If

so, we look at the series $\sum\limits_{k=1}^{\infty}|a_k|$ which makes every term positive.

If that series converges, then the original series $\sum\limits_{k=1}^{\infty}a_k$ must

converge as well. When both $\sum\limits_{k=1}^{\infty}a_k$ and $\sum\limits_{k=1}^{\infty}|a_k|$ converge, we

say that $\sum\limits_{k=1}^{\infty}a_k$ converges absolutely. If $\sum\limits_{k=1}^{\infty}|a_k|$ coverges and $\sum\limits_{k=1}^{\infty}a_k$

diverges, then we say that $\sum\limits_{k=1}^{\infty}a_k$ converges conditionally. For

example, $\sum\limits_{k=1}^{\infty}\frac{(-1)^k}{k^2}$ converges absolutely because $\sum\limits_{k=1}^{\infty}\left|\frac{(-1)^k}{k^2}\right|$, or,

equivalently, $\sum\limits_{k=1}^{\infty}\frac{1}{k^2}$ converges (*p*-series with $p>1$).

1. Divergence Test

   If $\lim\limits_{k\to\infty} a_k \neq 0$, the series $\sum\limits_{k=1}^{\infty} a_k$ diverges. It is impossible for a series to converge if we are not adding smaller and smaller numbers to it. The contrapositive of this statement, which is logically equivalent to the statement, is also very useful.

   That is, if $\sum\limits_{k=1}^{\infty} a_k$ converges, then $\lim\limits_{k\to\infty} a_k = 0$. In other words, if a series is convergent, its terms must approach zero. However, $\lim\limits_{k\to\infty} a_k = 0$ does not imply convergence.

   i. The series $\sum\limits_{k=1}^{\infty} \dfrac{k}{\sqrt{k^2+1}}$ is divergent since $\lim\limits_{k\to\infty} \dfrac{k}{\sqrt{k^2+1}} = \lim\limits_{k\to\infty} \dfrac{1}{\sqrt{1+ \frac{1}{k^2}}} = 1 \neq 0.$

   ii. An example of a series in which the terms approach zero but which is not convergent is the harmonic series, $\sum\limits_{k=1}^{\infty} \dfrac{1}{k}$.

2. Ratio Test

   (a) If $\lim\limits_{k\to\infty} \dfrac{a_{k+1}}{a_k} < 1$ then the series $\sum\limits_{k=1}^{\infty} a_k$ converges; (b) if $\lim\limits_{k\to\infty} \dfrac{a_{k+1}}{a_k} > 1$ the series diverges. If $\lim\limits_{k\to\infty} \dfrac{a_{k+1}}{a_k} = 1$, this test is inconclusive; use a different convergence test. Specifically, the Ratio Test does not work for p-series because in that case, $\lim\limits_{k\to\infty} \dfrac{a_{k+1}}{a_k} = 1$. Use this test mainly when a_k involves factorials or kth powers.

   For example, the series $\sum\limits_{k=1}^{\infty} \dfrac{1}{k!}$ converges since

   $$\lim\limits_{k\to\infty} \dfrac{\frac{1}{(k+1)!}}{\frac{1}{k!}} = \lim\limits_{k\to\infty} \dfrac{k!}{(k+1)!} = \lim\limits_{k\to\infty} \dfrac{k!}{(k+1)k!} = \lim\limits_{k\to\infty} \dfrac{1}{k+1} = 0 < 1.$$

Also, $\displaystyle\sum_{k=1}^{\infty}\frac{k^k}{k!}$ diverges because $\displaystyle\lim_{k\to\infty}\dfrac{\dfrac{(k+1)^{k+1}}{(k+1)!}}{\dfrac{k^k}{k!}}=$

$\displaystyle\lim_{k\to\infty}\frac{(k+1)^{k+1}}{(k+1)!}\cdot\frac{k!}{k^k}=\lim_{k\to\infty}\frac{(k+1)^k}{k^k}=\lim_{k\to\infty}\left(\frac{k+1}{k}\right)^k=$

$\displaystyle\lim_{k\to\infty}\left(1+\frac{1}{k}\right)^k=e>1.$

However, a much easier test is the divergence test:

$\displaystyle\lim_{k\to\infty}\frac{k^k}{k!}=\infty$ as the numerator will always be greater than

the denominator as they both have k terms but the numerator's factors are always k and the denominator's factors start with k and decrease down to 1.

3. Ratio test for absolute convergence

   (a) If $\displaystyle\lim_{k\to\infty}\left|\frac{a_{k+1}}{a_k}\right|<1$, then the series $\displaystyle\sum_{k=1}^{\infty}a_k$ converges; (b) if

   $\displaystyle\lim_{k\to\infty}\left|\frac{a_{k+1}}{a_k}\right|>1$, the series *diverges*. If $\displaystyle\lim_{k\to\infty}\left|\frac{a_{k+1}}{a_k}\right|=1$, this test is

   inconclusive, use a different convergence test. The series need not have positive terms and need not be alternating to use this test.

   i. The series $\displaystyle\sum_{k=1}^{\infty}\frac{(-1)^k 2^k}{k!}$ converges absolutely since

   $\displaystyle\lim_{k\to\infty}\frac{2^{k+1}}{(k+1)!}\cdot\frac{k!}{2^k}=\lim_{k\to\infty}\frac{2}{(k+1)}=0<1.$

   ii. The series $\displaystyle\sum_{k=1}^{\infty}\frac{(-1)^k}{k}$ does not converge absolutely (that is,

   $\displaystyle\sum_{k=1}^{\infty}\left|\frac{(-1)^k}{k}\right|$, which is equivalent to $\displaystyle\sum_{k=1}^{\infty}\frac{1}{k}$, does not converge

   because it is the harmonic series) but it converges without the absolute values by the alternating series test (the terms, in absolute value, decrease and approach zero). Thus,

   $\displaystyle\sum_{k=1}^{\infty}\frac{(-1)^k}{k}$ converges conditionally.

4. Direct Comparison Test

Suppose $\sum\limits_{k=1}^{\infty} a_k$ and $\sum\limits_{k=1}^{\infty} b_k$ are series with positive terms.

(a) If $\sum\limits_{k=1}^{\infty} b_k$ is convergent and $a_k \le b_k$ for all k, then $\sum\limits_{k=1}^{\infty} a_k$ converges. (b) If $\sum\limits_{k=1}^{\infty} b_k$ is divergent and $a_k \ge b_k$ for all k, then $\sum\limits_{k=1}^{\infty} a_k$ diverges. Part (a) says that if the series with larger terms converges, then the series with smaller terms converges. Part (b) says that if the series with smaller terms diverges, then the series with larger terms diverges. This test only applies to series with non-negative terms. Use this as a last resort, as other tests are often easier to apply.

i. For example, to see if $\sum\limits_{k=2}^{\infty} \dfrac{3k}{k^2-2}$ converges, we compare it to a similar (and, in this case, smaller) series, $\sum\limits_{k=2}^{\infty} \dfrac{3k}{k^2}$. This series diverges because it is equivalent to $3\sum\limits_{k=2}^{\infty} \dfrac{1}{k}$, which is a divergent harmonic series. Since the smaller series diverges, the larger (original) series diverges.

ii. To see if $\sum\limits_{k=1}^{\infty} \dfrac{5k}{2k^3+k^2+1}$ converges, we compare it to a similar (and, in this case, larger) series, $\sum\limits_{k=1}^{\infty} \dfrac{5k}{2k^3}$. This series converges because it is equivalent to a convergent p-series, $\dfrac{5}{2}\sum\limits_{k=1}^{\infty} \dfrac{k}{k^3} = \dfrac{5}{2}\sum\limits_{k=1}^{\infty} \dfrac{1}{k^2}$. Since the larger series converges, the smaller (original) series converges.

5. Limit Comparison Test

Suppose $\sum\limits_{k=1}^{\infty} a_k$ and $\sum\limits_{k=1}^{\infty} b_k$ are series with positive terms. If $\lim\limits_{k\to\infty} \dfrac{a_k}{b_k} = c$ where $0 < c < \infty$, then either both series converge or both series diverge. The limit comparison test chooses a series B whose convergence or divergence is known. It then compares another series A to B. If the ratio of the general terms gives a limit that is positive and finite, series A will have the same convergence/divergence result as B.

i. To see if $\sum\limits_{k=1}^{\infty} \dfrac{\sqrt{k}}{k^2+k+3}$ converges, compare it to

$$\sum\limits_{k=1}^{\infty} \dfrac{\sqrt{k}}{k^2} = \sum\limits_{k=1}^{\infty} \dfrac{1}{k^{3/2}},$$ which is a convergent *p*-series.

$$\lim_{k\to\infty} \dfrac{\dfrac{\sqrt{k}}{k^2+k+3}}{\dfrac{\sqrt{k}}{k^2}} = \lim_{k\to\infty} \dfrac{k^2}{k^2+k+3} = 1.$$ Since $0 < 1 < \infty$ and the

second series converges, then the first series also converges.

ii. To see if $\sum\limits_{n=1}^{\infty} \dfrac{10}{1+\sqrt{n}}$ converges, compare it to the divergent

p-series $\sum\limits_{n=1}^{\infty} \dfrac{1}{\sqrt{n}}$. $\lim\limits_{n\to\infty}\left(\dfrac{10}{1+\sqrt{n}} \cdot \dfrac{\sqrt{n}}{1} \right) = 10.$ Since the limit

exists, then $\sum\limits_{n=1}^{\infty} \dfrac{10}{1+\sqrt{n}}$ is also divergent.

6. Alternating Series Test

   If the alternating series $\sum\limits_{k=1}^{\infty} (-1)^k a_k = -a_1 + a_2 - a_3 + a_4 - a_5$
   $+ a_6 - \dots$ where $a_k > 0$ for all k, satisfies (a) $a_k > a_{k+1}$ <u>and</u>
   (b) $\lim\limits_{k\to\infty} a_k = 0$, then the series converges. If one of these
   conditions is not satisfied, the series diverges. That is, if
   each term is smaller than the previous term (in absolute
   value), and the terms are approaching zero, then the
   series converges. This applies only to alternating series.
   <u>Remainder:</u> $|R_k| \le a_{k+1}$. That is, when adding the first *n* terms
   of an alternating series, the remainder (or the error) is less
   than or equal to the first omitted term.

   $$\sum\limits_{k=1}^{\infty} \dfrac{(-1)^k}{k}$$ converges because, (a) $\dfrac{1}{k} > \dfrac{1}{k+1}$ for all k, and

   (b) $\lim\limits_{k\to\infty} \dfrac{1}{k} = 0.$

i. Find the number of terms required to approximate the sum of the series $\sum\limits_{n=1}^{\infty}(\cos n\pi)\dfrac{1}{n!}$ with an error of less than 0.001.

$\sum\limits_{n=1}^{\infty}(\cos n\pi)\dfrac{1}{n!} = -\dfrac{1}{1} + \dfrac{1}{2} - \dfrac{1}{6} + \dots$. This is an alternating series whose terms are getting smaller. The first factorial greater than 1,000 is 7!, so $\dfrac{1}{7!} < 0.001$. Therefore, it takes 6 terms to calculate $\sum\limits_{n=1}^{\infty}(\cos n\pi)\dfrac{1}{n!}$ with an error of 0.001.

7. **Integral Test**
Let $f(x)$ be a continuous, positive, decreasing function on $[1, \infty)$ which results when k is replaced by x in the formula for a_k. Then the series $\sum\limits_{k=1}^{\infty}a_k$ converges if and only if the improper integral $\int\limits_{1}^{\infty}f(x)dx$ converges.

Use this test when $f(x)$ is easy to integrate. This test only applies to series with positive terms.

$\sum\limits_{k=1}^{\infty}\dfrac{1}{k}$ diverges because $\int\limits_{1}^{\infty}\dfrac{1}{x}dx = \lim\limits_{x\to\infty}(\ln(x))\Big|_{1}^{\infty} = \infty$. That is, since the integral diverges, so does the series. Of course, we knew that because this series is harmonic and thus, divergent.

Keep in Mind...

➤ Do not confuse sequences with series—a series is the sum of the terms of a sequence.

➤ A series which converges absolutely, converges. This is confusing, but all it means is that if the series of absolute values of the terms converges, then the series itself converges. That is, if $\sum |u_k|$ converges, so does $\sum u_k$.

➤ Not every telescoping series converges!

➤ The divergence test is always the best place to start when deciding on whether a series is convergent. If $\lim\limits_{n\to\infty} a_n \neq 0$, then the series in question is divergent and no other tests are necessary. If $\lim\limits_{n\to\infty} a_n = 0$, then you go through all the other tests until you find one that is conclusive.

III. POWER SERIES

1. For a power series $\sum\limits_{k=0}^{\infty} c_k(x-a)^k$ exactly one of the following holds:

   (a) The series converges only for $x = a$.

   (b) The series converges absolutely (and hence converges) for all real values of x.

   (c) The series converges absolutely (and hence converges) for all x in some finite open interval $(a - R, a + R)$ and diverges if $x < a - R$ or $x > a + R$.

   At either of the points $x = a - R$ or $x = a + R$, the series may converge absolutely, converge conditionally, or diverge. The interval $(a - R, a + R)$ is called the interval of convergence and half of its size is called the radius of convergence of the series.

   The radius of convergence is represented by $R = \lim\limits_{n\to\infty} \left| \dfrac{c_{n+1}}{c_n} \right|$. The center of the series is $x = a$. To find the interval of convergence, use the Ratio Test for Absolute Convergence.

   i. For example, to find the interval of convergence of $\sum\limits_{k=1}^{\infty} \dfrac{(x-5)^k}{k^2}$, we apply the Ratio Test for Absolute Convergence: $\lim\limits_{k\to\infty} \left| \dfrac{a_{k+1}}{a_k} \right| = \lim\limits_{k\to\infty} \left| \dfrac{(x-5)^{k+1}}{(k+1)^2} \cdot \dfrac{k^2}{(x-5)^k} \right| = \lim\limits_{k\to\infty} |x - 5|$.

   For the series to converge, $|x - 5| < 1$. That is, $-1 < x - 5 < 1$, which implies that $4 < x < 6$. Remember that the series might or might not be convergent at the endpoints, so this must be checked by substituting each endpoint into the original series. For $x = 4$, the series becomes $\sum_{k=1}^{\infty} \frac{(-1)^k}{k^2}$ which converges absolutely and therefore converges. We know this because it is an alternating p-series with $p > 0$; also, $\sum_{k=1}^{\infty} \left| \frac{(-1)^k}{k^2} \right|$ is equivalent to the convergent p-series ($p > 1$) $\sum_{k=1}^{\infty} \frac{1}{k^2}$.

For $x = 6$, the series becomes $\sum_{k=1}^{\infty} \frac{1}{k^2}$ which is a convergent p-series ($p > 1$). So, the interval of convergence for $\sum_{k=1}^{\infty} \frac{(x-5)^k}{k^2}$ is [4, 6] and the radius of convergence is 1.

ii. To find the interval of convergence of $\sum_{k=0}^{\infty} k! x^k$ we apply the Ratio Test for Absolute Convergence:

$$\lim_{k \to \infty} \left| \frac{a_{k+1}}{a_k} \right| = \lim_{k \to \infty} \left| \frac{(k+1)! x^{k+1}}{k! x^k} \right| = \lim_{k \to \infty} (k+1) |x| = \infty.$$ This means that

the series diverges for all nonzero values of x. So it converges only at $x = 0$, and the radius of convergence is $R = 0$.

IV. TAYLOR SERIES

A. One can use the Taylor Series to approximate a function, $f(x)$, by a polynomial of specified degree in the vicinity of a given point, $x = a$. This is an extension of finding the equation of the tangent line to a function at a point. Just like the tangent line approximates the function fairly well in a small vicinity of the tangency point, the Taylor polynomial approximates the function, $f(x)$, in the vicinity of $x = a$. The difference between approximating the values of a function using a tangent line at a point and using a Taylor polynomial, is that the Taylor polynomial can be more accurate in a larger vicinity of the point. The higher the degree of the polynomial, the better the approximation.

1. If f has derivatives of all orders at $x = a$, then the Taylor series

   for f about $x = a$ is given by: $f(n) = \sum\limits_{n=0}^{\infty} \dfrac{f^{(n)}(a)}{n!}(x-a)^n = f(a) +$

   $f'(a)(x-a) + \dfrac{f''(a)}{2!}(x-a)^2 + \dfrac{f'''(a)}{3!}(x-a)^3 + \cdots + \dfrac{f^n(a)}{n!}(x-a)^n + \ldots$.

   In the special case in which $a = 0$, the Taylor

   series is called the Maclaurin series for f.

   In that case, $\sum\limits_{k=0}^{\infty} \dfrac{f^{(n)}(a)}{n!}(x)^n = f(0) +$

   $f'(0)(x) + \dfrac{f''(0)}{2!}(x)^2 + \dfrac{f'''(0)}{3!}(x)^3 + \cdots + \dfrac{f^n(0)}{n!}(x)^n + \ldots$.

   i. For example, find the Taylor series of degree 3 for $f(x) = \dfrac{1}{x}$

      about $x = 1$. First find the derivatives of f: $f'(x) = -\dfrac{1}{x^2}$,

      $f''(x) = \dfrac{2}{x^3}$, $f'''(x) = -\dfrac{6}{x^4}$. Next, evaluate the function and

      its derivatives at $x = 1$: $f(1) = 1$, $f'(1) = -1$, $f''(1) = 2 = 2!$,
      $f'''(1) = -6 = -3!$. Now substitute all the values into the
      Taylor series above to get: $P_3(x) = 1 - (x-1) + (x-1)^2 - (x-1)^3$. As seen in the graph below, the Taylor polynomial
      approximates $f(x)$ in the vicinity of $x = 1$.

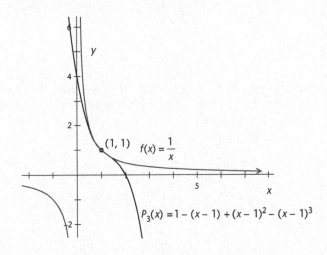

ii. Find the Maclaurin series of degree four for $f(x) = e^x$.
A Maclaurin polynomial is a Taylor polynomial centered
at $x = 0$. Since $f(x) = f'(x) = f''(x) = f'''(x) = f^{(4)}(x)$, it follows
that $f(0) = f'(0) = f''(0) = f'''(0) = f^{(4)}(0) = 1$.

Substituting into the Maclaurin series, we have:

$$P_4(x) = 1 + x + \frac{x^2}{2!} + \frac{x^3}{3!} + \frac{x^4}{4!}.$$

iii. Common Maclaurin series (and their intervals of
convergence) that must be memorized:

$$\frac{1}{1-x} = 1 + x + x^2 + x^3 + \cdots = \sum_{k=0}^{\infty} x^k \text{ for } -1 < x < 1$$

$$e^x = 1 + x + \frac{x^2}{2!} + \frac{x^3}{3!} + \frac{x^4}{4!} + \cdots = \sum_{k=0}^{\infty} \frac{x^k}{k!} \text{ for } -\infty < x < \infty$$

$$\sin(x) = x - \frac{x^3}{3!} + \frac{x^5}{5!} - \frac{x^7}{7!} + \cdots = \sum_{k=0}^{\infty} (-1)^k \frac{x^{2k+1}}{(2k+1)!} \text{ for } -\infty < x < \infty$$

$$\cos(x) = 1 - \frac{x^2}{2!} + \frac{x^4}{4!} - \frac{x^6}{6!} + \cdots = \sum_{k=0}^{\infty} (-1)^k \frac{x^{2k}}{(2k)!} \text{ for } -\infty < x < \infty$$

$$\ln(1+x) = x - \frac{x^2}{2} + \frac{x^3}{3} - \frac{x^4}{4} + \cdots = \sum_{k=0}^{\infty} (-1)^k \frac{x^{k+1}}{(k+1)} \text{ for } -1 < x \leq 1$$

$$\tan^{-1}(x) = x - \frac{x^3}{3} + \frac{x^5}{5} - \frac{x^7}{7} + \cdots = \sum_{k=0}^{\infty} (-1)^k \frac{x^{2k+1}}{2k+1} \text{ for } -1 \leq x \leq 1.$$

2. Creating new power series from known power series. This
is done by substituting x with a different quantity—in
the known series expansion as well as in the interval of
convergence!

i. For instance, a Maclaurin series for e^{-x^2} can be obtained
by substituting $-x^2$ for x in the Maclaurin series for e^x. So,

$$e^{-x^2} = 1 + (-x^2) + \frac{(-x^2)^2}{2!} + \frac{(-x^2)^3}{3!} + \frac{(-x^2)^4}{4!} + \cdots =$$

$1 - x^2 + \frac{x^4}{2!} - \frac{x^6}{3!} + \frac{x^8}{4!} + \cdots$. You may be asked to determine the general term of the power series which is $\displaystyle\sum_{n=0}^{\infty}\frac{(-1)^n x^{2n}}{n!}$.

Substituting x with $-x^2$ in the interval of convergence for e^x, yields $-\infty < -x^2 < \infty \rightarrow 0 < x^2 < \infty \rightarrow -\infty < x < \infty$.

ii. Also, we can write a Maclaurin series for $\dfrac{x}{1+x^3}$

by substituting $-x^3$ for x in the Maclaurin series

for $\dfrac{1}{1-x}$ and then multiplying the series by x:

$$\frac{1}{1+x^3} = 1 + (-x^3) + (-x^3)^2 + (-x^3)^3 + \cdots = 1 - x^3 + x^6 - x^9 + \cdots.$$

Multiplying by x, we get $\dfrac{x}{1+x^3} = x - x^4 + x^7 - x^{10} + \cdots$. The

general term is $\displaystyle\sum_{n=0}^{\infty}(-1)^n x^{3n+1}$. Substituting $-x^3$ into the

interval of convergence for $\dfrac{1}{1-x}$, yields $-1 < -x^3 < 1 \rightarrow$
$-1 < x < 1$.

iii. The power series for $\tan^{-1}(2x)$ can be obtained by substituting x with $2x$ in the power series for $\tan^{-1}(x)$:

$$\tan^{-1}(2x) = 2x - \frac{(2x)^3}{3} + \frac{(2x)^5}{5} - \frac{(2x)^7}{7} + \cdots =$$

$2x - \dfrac{8x^3}{3} + \dfrac{32x^5}{5} - \dfrac{128x^7}{7} + \cdots$. The general term is

$\displaystyle\sum_{n=0}^{\infty}\frac{(-1)^n (2x)^{2n+1}}{2n+1}$. Substituting x with $2x$ in the interval of

convergence of $\tan^{-1}(x)$ yields: $-1 \le 2x \le 1 \rightarrow -\dfrac{1}{2} \le x \le \dfrac{1}{2}$.

3. Differentiating and Integrating power series is done term by term. The interval and radius of convergence remain unchanged.

    i. For instance, to show that $\dfrac{d}{dx}(\sin(x)) = \cos(x)$, we

    differentiate the Maclaurin series for $\sin(x)$ term by term:

    $$\frac{d}{dx}(\sin(x)) = \frac{d}{dx}\left(x - \frac{x^3}{3!} + \frac{x^5}{5!} - \frac{x^7}{7!} + \cdots\right) = 1 - \frac{x^2}{2!} + \frac{x^4}{4!} - \frac{x^6}{6!} + \cdots =$$

    $\cos(x)$.

ii. To show that $\int \dfrac{1}{1+x^2}dx = \tan^{-1}(x)+C$, integrate the

Maclaurin series for $\dfrac{1}{1+x^2}$. We must first create this series

by substituting x with $-x^2$ in the Maclaurin series for $\dfrac{1}{1-x}$:

$$\dfrac{1}{1+x^2} = 1+(-x^2)+(-x^2)^2+(-x^2)^3+\cdots = 1-x^2+x^4-x^6+\cdots.$$

Integrating this series term by term yields: $x-\dfrac{x^3}{3}+\dfrac{x^5}{5}-$

$\dfrac{x^7}{7}+\cdots+C = \tan^{-1}(x)+C$ since $\tan^{-1}(0) = 0$, $C = 0$. So,

$\tan^{-1}(x) = x-\dfrac{x^3}{3}+\dfrac{x^5}{5}-\dfrac{x^7}{7}+\cdots.$

4. Lagrange's form of the remainder: $R_n(x) = \dfrac{f^{(n+1)}(c)(x-a)^{n+1}}{(n+1)!}$

where c is between a and x. This represents the error
that arises when approximating a function with a Taylor
polynomial of degree n. This is used to prove that a certain
function is approximated by a series for all x-values.

i. To show that the Maclaurin series for $\cos(x)$ converges for
all values of x, we must show that $\lim\limits_{x\to\infty} R_n(x) = 0$. Now,
$f^{(n+1)}(x) = \pm\cos(x)$ or $f^{(n+1)}(x) = \pm\sin(x)$. In either case,
for any value of c, $f^{(n+1)}(c) \leq 1$. In this case, $a = 0$ so

$$|R_n(x)| = \left|\dfrac{f^{(n+1)}(x-a)^{n+1}}{(n+1)!}\right| \leq \dfrac{|x|^{n+1}}{(n+1)!} \text{ approaches 0 as}$$

n approaches ∞.

ii. Lagrange's form of the remainder, $R_n(x) = \dfrac{f^{(n+1)}(c)(x-a)^{n+1}}{(n+1)!}$,

is also used when performing computations using Taylor
series. For instance, suppose we try to approximate
$\sin\left(\dfrac{\pi}{3}\right)$ such that $R_n(x) \leq 0.000005$. We'll use the
Maclaurin series for $\sin(x)$, hence, $a = 0$. Since
$f^{(n+1)}(x) = \pm\cos(x)$ or $f^{(n+1)}(x) = \pm\sin(x)$, $f^{(n+1)}(x) \leq 1$.

Thus, $R_n(x) = \left|\dfrac{f^{(n+1)}(c)(x-a)^{n+1}}{(n+1)!}\right| \leq \dfrac{|x|^{n+1}}{(n+1)!}$.

Since $x = \dfrac{\pi}{3}$, we must find an n value that satisfies

$$\left| R_n\left(\frac{\pi}{3}\right) \right| \le \frac{\left|\frac{\pi}{3}\right|^{n+1}}{(n+1)!} < 0.000005.$$ By trial and error, $n = 8$.

Therefore, $\sin\left(\dfrac{\pi}{3}\right) = \dfrac{\pi}{3} - \dfrac{\left(\frac{\pi}{3}\right)^3}{3!} + \dfrac{\left(\frac{\pi}{3}\right)^5}{5!} - \dfrac{\left(\frac{\pi}{3}\right)^7}{7!} = .8660212717.$

Using the calculator, $\sin\left(\dfrac{\pi}{3}\right) = .8660254038.$

Keep in Mind...

➤ Memorize the basic Maclaurin series because you will not have time to derive them on the AP® exam.

➤ Apply the ratio test for absolute convergence to a series when finding the interval or radius of convergence.

➤ Don't forget to test the endpoints of the interval of convergence to see if the series is convergent or not at these points.

➤ There is a difference between a Taylor series and a Taylor polynomial. The Taylor series is infinite while the Taylor polynomial has a finite number of terms.

CHAPTER 11
PRACTICE PROBLEMS

(See solutions on page 239)

1. $\lim\limits_{n \to \infty} \left\{ \dfrac{2n-1}{3en^2 - 1} \right\} =$

2. State whether or not the series converges and name the test used:

   (A) $\sum\limits_{k=1}^{\infty} \dfrac{\cos k}{k^2}$

   (B) $\sum\limits_{k=1}^{\infty} \dfrac{e^k}{k+1}$

   (C) $\sum\limits_{k=1}^{\infty} \dfrac{(-1)^k}{(k-1)!}$

3. What is the difference between a Taylor polynomial and a Maclaurin polynomial?

4. Find the interval and radius of convergence of $\sum\limits_{k=1}^{\infty} \dfrac{(2-x)^k}{k^3}$.

5. Find a Taylor series of degree four for $y = 3e^x$ at $x = 1$.

6. Find a Maclaurin series of degree six for $y = \dfrac{\sin(2x)}{x}$.

PART VI
THE EXAM

The Graphing Calculator

I. INTRODUCTION

A. Several brands and models of calculators are helpful and necessary for the AP® Calculus exam. Calculators from Casio, Hewlett-Packard, and Sharp have been approved by the College Board but by far, the biggest provider of calculators is Texas Instruments.

While TI has several newer calculators on the market that have color, are much more efficient with memory, and are noticeably faster, the technology has not changed much for years. The screen interface is easy to use with relatively few keystrokes to get the needed results. Despite their age, TI models remain the calculator of choice. Most books that incorporate calculator technology into the teaching of calculus use images from the TI-82, TI-83, TI-84, and TI-84 Plus calculators, and the following sections will do the same. The newer calculators still use the same keyboard and keystrokes as their predecessors.

The calculator can perform several tasks, and the designers of the AP® exam go to great lengths to be sure that it is a tool that can help students but not solve actual problems. If there is a calculator function on the approved list of calculators that can actually solve a problem, that problem would not appear on the section(s) where calculators are permitted. For example, if a student were asked to find the derivative of x^5, it would not appear on the calculator section of the exam because any calculator with a computer algebra system built in like the TI-89 and TI-NSpire CAS, which are approved, can find it symbolically. No student is allowed to gain an advantage based on the type of calculator selected.

B. Other than being used as a scientific calculator (calculations using addition, subtraction, multiplication, division, exponentiation, roots, trig and logs), calculators can be used for only 4 specific applications:

1. plotting the graph of a function within an arbitrary viewing window
   - included is evaluating the function at a point
2. finding the zeros of functions (solving equations numerically)
   - included is finding intersections of two functions
3. finding the numerical value of a derivative
   - this means finding the value of the derivative of a function at some value of x
4. numerically calculating the value of a definite integral

C. Following is a basic guide to the use of the calculator. For more information, go online for specific directions for your particular calculator. For purposes of this chapter, we will assume that students have a TI-83 or TI-84. More recent models have enhanced screens and color options that are not pertinent to the mathematics. For this purpose, the newer and older calculators are equivalent.

Students should absolutely tend to several matters before the test. First is the amount of accuracy necessary for answers. Standard in AP® exams is 3-decimal place accuracy. So, students should press the (MODE) button and change the second line from FLOAT to 3 using the right arrow button. This will ensure all answers will be represented to 3 decimal places (an answer of 7 will appear as 7.000).

The second matter is vitally important: The calculator must be set to **radian mode**, not degree mode. Sometimes students doing trigonometry problems change to degree mode and fail to change it back. Because of the techniques the calculator uses in taking derivatives and integrals, failure to change to radian mode, even in problems not involving trig, can be disastrous. For instance in the calculator section of the multiple choice, you might see the problem: $\int_0^2 x^2 \sin x \, dx$. This is easily performed on the calculator and the correct answer is 2.470. But one of the 4 choices would probably be 0.070, the result you would get if your calculator was in degree mode. You would confidently choose this answer not knowing that it was completely wrong because your calculator was not in radian mode. So, especially if you are using a calculator that is not yours, perform these two housekeeping chores when you first turn on the calculator.

II. PLOTTING GRAPH OF FUNCTION IN ARBITRARY VIEWING WINDOW AND EVALUATING FUNCTIONS AT A POINT

A. To graph $y = x^2 - 4$, press the ⓨ= button and enter the function into Y1. The variable x is generated using the (X,T,θ,n) button. Remember that there is a difference between the minus sign (–) and the negation sign at the bottom of the keyboard (-).

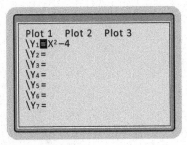

Pressing the (GRAPH) button should give you this result.

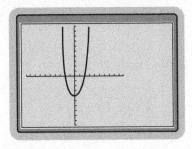

Realize that a graph can never be used to justify an answer as many AP® free-response questions ask students to do. For instance, if a student is asked to find intervals of increasing and decreasing for $y = x^2 - 4$, the graph itself cannot be used as a justification that the function increases on (0, ∞) and decreases on $(-\infty, 0)$. The first derivative test must be used. The graph can confirm but not justify your conclusion.

B. Evaluating a function at a point—that is, finding the *y*-value given an *x*-value. For instance, evaluate $f(3)$ given that $f(x) = 3e^x - x^2$.

1. Method 1. By brute force, you can replace *x* with 3:

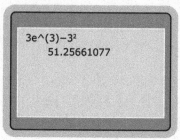

2. Method 2. You may store 3 into *x* and then type in $3e^x - x^2$. To do this, press ③ (STO →) (X) (ENTER) (2nd) (LN) (X) (⟩) (-) (LN) (X) (x²) (ENTER).

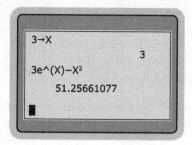

3. Method 3. Another way to substitute *x* with 3 is to use the CALCULATE menu to get:

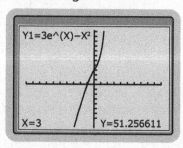

To do this, first enter the function in (Y=) then press (2nd) (Trace) ① ③ (ENTER).

Notice that the *y*-value in this case is rounded off to 6 decimal places instead of 8. However, this will not matter on the exam since you are required to use at least three-decimal-place accuracy. Also, when using this method, make sure the number you are substituting is between Xmin and Xmax; otherwise you will get an error message that looks like this:

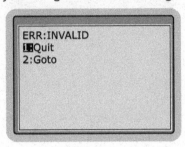

```
ERR:INVALID
1:Quit
2:Goto
```

For instance, if you wanted to evaluate *f*(13), but your window only contains *x* values from –10 to 10, you will get the above error message.

4. Method 4. You can evaluate *f*(3) by looking at the table of values. To do this, first enter the function in (Y₁), then press (2ⁿᵈ) (GRAPH) and scroll to *x* = 3.

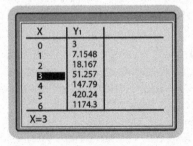

X	Y₁
0	3
1	7.1548
2	18.167
3	51.257
4	147.79
5	420.24
6	1174.3
X=3	

Notice that the *y*-value is rounded off to 3 decimal places— this is okay if this number is your final answer, but not if you need it in an earlier step. Remember, you are to round off— if you'd like to or if the problem asks you to—only at the very end of your calculations, otherwise you will accumulate round-off errors. You can also set the table of values to allow you to enter the *x*-value you want without scrolling. To do this, press (2ⁿᵈ) (WINDOW). Highlight *Ask* in the first line. After this, when you look at the table, it will be empty. This is because it is waiting for you to enter x-values.

5. Method 5. You can have the calculator substitute x with 3 for you:

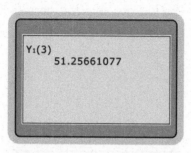

To do this, first enter the function in $(Y=)$, then press $(VARS)$ $(Y\text{-}VARS)$ ① ① ((③)) $(ENTER)$. (These steps assume that the function was entered in Y_1.) If the function was entered in Y_2 then the directions would be $(VARS)$ $(Y\text{-}VARS)$ ① ② ((③)) $(ENTER)$.

 ## III. FINDING THE ZEROS OF FUNCTIONS AND INTERSECTION OF 2 FUNCTIONS

A. Finding the zeros is also known as finding the roots or x-intercepts of the function. This feature will be used mostly to find critical values (roots of a derivative) or inflection points (roots of a second degree function).

To illustrate, find the x-value where $f(x) = 3e^x - \dfrac{x^3}{3}$ has a relative minimum and justify your answer.

1. The calculator has the ability to find relative minimum or relative maximum values of functions in an interval (use the $(2ND)$ $(TRACE)$ menu). But this is a matter of key-pressing and the AP® exam is about showing your calculus knowledge. So, we will use the first-derivative test. We first find the derivative: $f'(x) = 3e^x - x^2$.

2. Enter the function in (Y=) then press (2nd) (TRACE) (2). The calculator will ask you to enter a left bound. This is an *x* value less than the root. In this case, it looks like the root is somewhere between 0 and –2. So press (−2) and (ENTER) for the left bound.

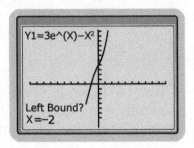

When you are asked for a right bound, that is, a number greater than the root, press (0) and (ENTER).

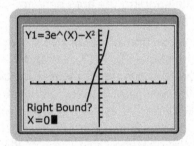

You are next asked to guess. Disregard this; just press (ENTER).

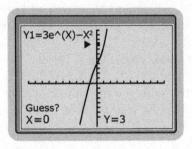

The answer is below:

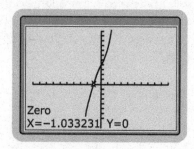

Zero
X=-1.033231 Y=0

3. Since $f'(x) < 0$ on $(-\infty, 1.033)$ and $f'(x) > 0$ on $(1.033, \infty)$, $f(x)$ has a relative minimum and absolute minimum at $x = -1.033$. If you want to find the absolute minimum value of $f(x)$, use the calculator to find $f(-1.033) = 3e^{-1.033} - \frac{(-1.033)^3}{3} = 1.435$.

B. Finding the intersection points of two functions. This is used most often when finding the area between two curves or volumes of solids of revolution.

1. Find the intersection points of $f(x) = x^2$ and $g(x) = \sin(x)$. Enter the functions in (Y=) ($Y_1 = x^2$ and $Y_2 = \sin(x)$). Change the window to trig window by pressing (ZOOM) (7) and make sure that the calculator is in radian mode (click (MODE) and highlight Radian—99.9% of the time, your calculator needs to be in radian mode for the AP® exam). Press (2nd) (TRACE) (5). Disregard the question "First Curve?" Place the cursor as close to the intersection you are looking for as possible, and press (ENTER). Disregard the question "Second Curve" and press (ENTER). Disregard "Guess?" and press (ENTER). The given functions intersect at (0, 0) and (.87672622, .76864886). For short, after pressing (2nd) (TRACE) (5) and placing the cursor on the intersection, press (ENTER) three times to get the answer. Make sure you see the word INTERSECTION above the answer; otherwise you are not done. The graphs that follow are zoomed in for clarity's sake.

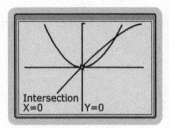
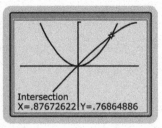

IV. FINDING THE LIMIT OF A FUNCTION

A. To evaluate $\lim_{x \to a} f(x)$ you must replace x with values that are very close to a, from both sides of a. To evaluate $\lim_{x \to \pm\infty} f(x)$, you must replace x with very large values (if $x \to \infty$) or very low values (if $x \to -\infty$).

   1. For instance, to evaluate $\lim_{x \to 0} \frac{1}{x}$, enter $y = \frac{1}{x}$ into Y_1 and then evaluate it with values of x that approach zero from the left of zero:

```
Y₁ (−.01)
                    −100
Y₁ (−.001)
                    −1000
Y₁ (−.0001)
                    −10000
```

From here we see that $\lim_{x \to 0^-} \frac{1}{x} = -\infty$. Checking the right-sided limit, we get:

```
Y₁ (.01)
                    100
Y₁ (.001)
                    1000
Y₁ (.0001)
                    10000
```

This is good evidence that $\lim\limits_{x\to 0^+} \dfrac{1}{x} = +\infty$. Since the left and right-hand limits are not equal, we conclude that $\lim\limits_{x\to 0} \dfrac{1}{x}$ does not exist.

2. To evaluate $\lim\limits_{x\to\infty} \dfrac{1}{x}$, enter $y = \dfrac{1}{x}$ into Y₁, substitute x with large values:

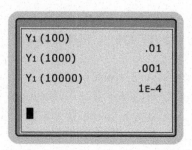

These calculations show that $\lim\limits_{x\to\infty} \dfrac{1}{x} = 0$. Note that you can press (2nd) (ENTER) after the first calculation to save time. This will copy your last step so all you need is to replace the 100 with 1000.

3. The limit of a function can also be evaluated by looking at its graph. From this graph we can see that $x = 0$ is a vertical asymptote of $y = \dfrac{1}{x}$ so $\lim\limits_{x\to 0} \dfrac{1}{x}$ DNE.

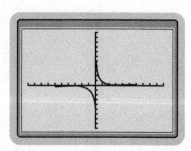

Test Tip

Because of its nature, an approved AP® exam graphing calculator cannot actually find limits but can approximate them. Therefore, any question concerning limits must be shown by analytic methods. Still, the calculator can be used to verify your answers using the techniques shown in this section. If you encounter a limit question and a calculator is permitted, you might wish to verify your answer using the calculator. But in the exam, that isn't proof, so don't mention it in your written answer.

V. EVALUATING DERIVATIVES AND DRAWING THEIR GRAPHS

A. There are at least four ways to evaluate the derivative of a function at a given x-value. For instance, find $f'(-2)$ if $f(x) = \ln(1 - x)$.

1. You could find the derivative on your own and enter it in Y_1. That is, since $f'(x) = -\dfrac{1}{1 - x}$ set $Y_1 = -\dfrac{1}{1 - x}$ and, on the home screen, press (VARS) (YVARS) (1) (1) (() (-2) ()) (ENTER).

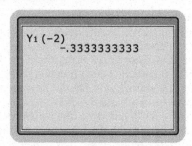

2. Or, you can let the calculator find the derivative and substitute in $x = -2$. In (Y=) enter $Y_1 = \ln(1 - x)$. On the home screen (press (2nd) (Mode) to quit any screen you're in and go to the home screen), press (MATH) (8) (VARS) (YVARS) (1) (1) (,) (X) (,) (-2) ()) (ENTER). (Note that you must press the comma button after Y_1 and after X.)

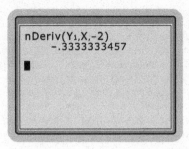

Note that this answer is not equivalent to the actual answer, $-\dfrac{1}{3}$, but it is a good enough approximation for the exam.

The more recent TI calculator models have a feature, MathPrint. It allows users to see calculator inputs and output the way it appears in textbooks. So when using MathPrint, 1/2 will appear as $\dfrac{1}{2}$, 2^4 will appear as 2^4, 2^(1/2) will

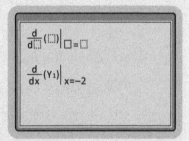

appear as $\sqrt{2}$ as well as other changes. One big change is that the nDeriv function when used as shown, it will appear as in the screenshot to the right. There are 3 inputs: the variable of interest (usually x), the function (either written directly or where it is stored), and finally the value of the variable where you want to evaluate the derivative. So, for the function $y = \ln(1 - x)$ stored in Y1, the statement below will generate $y'(-2)$ when MathPrint is enabled. MathPrint can be turned on using the Mode screen and scrolling down or pressing (2ND) (MODE). Newer calculators have MathPrint as the default although Classic is available. But the math is the same.

3. You can also enter (MATH) (8) (LN) (1) (−) (X) (⟩) (,) (X) (,) (-2) (⟩) (ENTER) to get the answer.

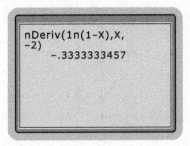

```
nDeriv(1n(1-X),X,
-2)
          -.3333333457
```

4. Once again, after graphing the function, you can use the CALCULATE menu. Press (2nd) (TRACE) (6) (-2) (ENTER).

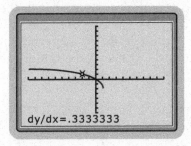

```
dy/dx=.3333333
```

Again, make sure that the x-value you are entering is in your window, or else you will get an error message. In this case, any x value less than –10 or greater than 10 would produce an error. Also, make sure that the x-value you enter is in the domain of the function. You will not get an error, but no answer to your query. For instance, asking for the derivative of the function above at x = 3 would simply give the output X = 3 Y = . There is no derivative given because the function (and thus the derivative) does not exist at x = 3.

B. Drawing the graph of f'(x) and f"(x).

1. To draw the graph of f'(x), press (Y=) and enter f(x) in Y₁. Then in Y₂ enter (MATH) (8) (VARS) (YVARS) (1) (1) (,) (X) (,) (X) (1) (ENTER). For instance, graph the derivative of f(x) = x⁴.

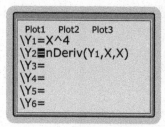
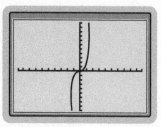

Notice that if you want to draw only the derivative graph, Y_2, you must disable the function graph, Y_1. (To disable an equation, place the cursor on the equal sign and press (ENTER). The equation will remain, but the calculator will not graph it. To enable the function, place the cursor on the equal sign and press (ENTER).) You can have both equations graphed at once, or even better, highlight one of them to tell the difference more easily:

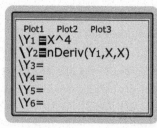
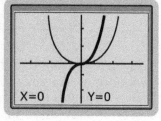

To highlight a function, place the cursor to the left of Y_2 and press (ENTER). To change it back, press (ENTER) 6 times. In this case, the derivative was highlighted. When graphing the above functions, the zoom-in feature was used for clarity.

2. To draw the graph of $f''(x)$ (given that the functions are entered as above in Y_1 and Y_2) enter the following in Y_3: (MATH) (8) (VARS) (YVARS) (1) (2) (,) (X) (,) (X) (J) (ENTER)). The graph of the second derivative is dotted.

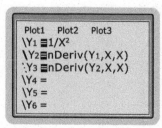
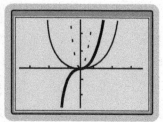

BIG DEAL: The calculator is by no means a perfect tool. Here are some instances to watch out for:

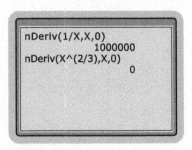

```
nDeriv(1/X,X,0)
                  1000000
nDeriv(X^(2/3),X,0)
                        0
```

In the first instance, the calculator states that $f'(0) = 1000000$ for $f(x) = \dfrac{1}{x}$. We know this to be wrong since $f'(x) = -\dfrac{1}{x^2}$ and hence, $f'(0)$ does not exist.

In the second instance, the calculator states that $f'(0) = 0$ for $f(x) = x^{\frac{2}{3}}$. We know this not to be true because $f'(x) = \dfrac{2}{3x^{\frac{1}{3}}}$, and hence $f'(0)$ does not exist.

The reason for this is that the calculator does not actually find the derivative at 0 but approximates it by finding the slope of the line between points just to the left of $x = 0$ and to the right of $x = 0$. Normally, this will be a good approximation for the derivative, but in this case, because the curve is not differentiable at $x = 0$, this method gives a false result.

VI. EVALUATING DEFINITE INTEGRALS

A. There are three ways of evaluating definite integrals, which mostly occur when finding area or volume:

1. To evaluate $\int_{-2}^{3} x^2 dx$, on the home screen, enter (MATH) (9) (X) (X²) (,) (X) (,) (−2) (,) (3) (①) (ENTER). Note that you must press the comma button after X², X, and −2.

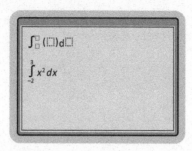

2. To evaluate $\int_{-2}^{3} x^2 dx$, enter x^2 in (Y=), more specifically in Y_1. Then, on the home screen, enter (MATH) (9) (VARS) (YVARS) (1) (1) (,) (X) (,) (-2) (,) (3) ()) (ENTER).

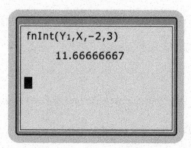

3. The third, more visual way to evaluate $\int_{-2}^{3} x^2 dx$, is to use the CALCULATE menu after having entered the function in (Y=). Press (2nd) (TRACE) (7) (-2) (3) (ENTER).

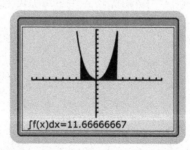

4. To evaluate an improper integral of the form $\int_{2}^{\infty} \frac{1}{x^2} dx$, let the upper limit get larger and larger by using the table of values.

In this case, the last x in the expression entered in Y_2 stands for the upper limit of the integral. If we allow this upper limit to get larger and larger, 10, 100, 1000, and so on, the integral's value approaches 0.5. Evaluate $\int_2^\infty \frac{1}{x^2} dx$ by hand to verify that $\int_2^\infty \frac{1}{x^2} dx = 0.5$.

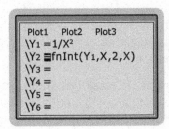
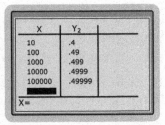

Note that the original function has been disabled so that the table only contains values of Y_2, those of the integral. Also, the table of values is in **Ask** mode for the independent variable.

CALCULATOR TIPS:

To stop the calculator in the middle of graphing a function or in the middle of a calculation that takes too long, press (ON).

To go back to the previous line on the home screen, press (2nd) (ENTER). The calculator can memorize about 25 steps! So if you need to evaluate a function at more than one x-value, instead of re-entering the function every time, just press (2nd) (ENTER) and then substitute the new x-value.

Don't confuse the subtraction key with the negative key. They are not interchangeable!

The CLEAR button clears a whole statement; the DELETE button erases one character at a time.

To go to the beginning of a statement, press (2nd) (←).
To go to the end of a statement, press (2nd) (→).

To draw a vertical line, say $x = 6$, you need to use the Draw menu. Press (2nd) (PRGM) (4) (6). To erase any drawing, press (2nd) (PRGM) (1).

You cannot evaluate, find roots, max/min points, intersection points, or derivatives of drawing objects.

To draw the tangent line to a function stored in Y_1 at a point, say $x = 3$, press (2nd) (PRGM) (5) (VARS) (Y-VARS) (1) (3) (ENTER). For instance, if you are asked to find the equation of the tangent line to a function at a point, you can graph the equation of the tangent line you've found and then draw it using the Draw menu to compare.

If you've made a mistake and need to erase a character and replace it with another, or if a character is missing, do not erase the whole statement; instead, use the insert feature, (2nd) (DEL). This creates an empty space for your new character.

To use the calculator with parametric equations, press (MODE) Par. Using the CALCULATE menu with parametric equations is very similar to using it with Cartesian equations. Make sure your t-step is small, about 0.1, so the graph will not be jagged.

To use the calculator with polar equations, press (MODE) Pol. The CALCULATE menu is also simple to use. Just make sure that your θ_{step} is small, about 0.1.

ERR:SYNTAX means you typed in something wrong—perhaps an extra comma, a subtraction sign instead of a negative sign, etc. Generally, if you press 2, the calculator will place the cursor on the error and you can correct it.

If you press (MODE) G-T, you can see the graph along with the table on the same screen! If you press (GRAPH), you can trace the curve and see the points in the table at the same time. If you press (2nd) (GRAPH), you can scroll down the x-values in the table.

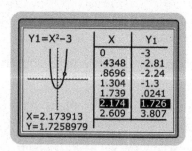

Practice Multiple-Choice Questions

Practice with the following AP®-style questions. Then go online to access our timed, full-length practice exam at *www.rea.com/studycenter*.

Problems 1–20 are AB problems and 21–25 are BC problems. Note that multiple concepts are used within each problem, making them slightly more complex. Graphing calculators should not be used other than on indicated problems.

1. Let $f(x) = x^3$. What is the difference between the average rate of change of f on the interval $[-2, 4]$ and the average value of $f(x)$ on the interval $[-2, 4]$?

    (A) 0

    (B) $\dfrac{2}{3}$

    (C) 2

    (D) 12

2. Let g be a function given by $g(x) = \sqrt{x}f(x)$. If $f(4) = 8$ and $f'(4) = -3$, use the tangent line to g at $x = 4$ to approximate $g(4.5)$.

    (A) 14

    (B) 13

    (C) $\dfrac{125}{8}$

    (D) −18

3. The graph of the function *f* on the interval [–6, 6] is shown below. Let *g* be the continuous function defined by $g(x) = \int_0^x f(t)\,dt$. For what value of *x* does *g* have its absolute minimum?

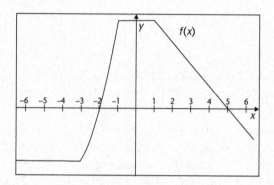

(A) $x = -3$

(B) $x = -2$

(C) $x = 5$

(D) $x = 6$

4. Find $\lim\limits_{x \to 1} \dfrac{x \sin \pi x - \sin \pi x - x + 1}{x \cos \pi x - \cos \pi x - x + 1}$

(A) 0

(B) –1

(C) $\dfrac{1}{2}$

(D) does not exist

5. The function $f(x)$ is continuous on the interval $[-4, 8]$. Selected values of x and $f(x)$ are given in the table below. $\int_{-4}^{8} f(x)\,dx$ is approximated with left Riemann sums with 4 equal subintervals, right Riemann sums with 4 equal subintervals, 4 trapezoids with 4 equal subintervals, and 2 midpoint rectangles with 2 equal subintervals. If $\int_{-4}^{8} f(x)\,dx = 30$, which approximation is closest?

x	-4	-1	2	5	8
$f(x)$	-6	-1	2	5	7

(A) Left Riemann Sum

(B) Right Riemann Sum

(C) Trapezoid

(D) Midpoint

6. If $f(x) = (3 - 2\sin x^2)^3$, find $f'(x)$

(A) $(3 - 2\sin x^2)^2$

(B) $-12x(3 - 2\sin x^2)^2$

(C) $-(3 - 2\sin x^2)^2(\cos x^2)$

(D) $-12x(3 - 2\sin x^2)^2(\cos x^2)$

7. Sunlight shining through a window creates a parallelogram-shaped shadow on the ground with sides *a* and *b* and angle θ as shown in the figure to the right. If the sides do not change but the angle of the shadow increases by $\dfrac{10°}{\text{min}}$, how fast is the area of the shadow changing when *a* = 10 ft, *b* = 8 ft, and θ = 120°? The area of a parallelogram is $A = ab\sin\theta$.

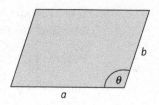

(A) Decreases by $\dfrac{4.2 \text{ ft}^2}{\text{min}}$

(B) Decreases by $\dfrac{24 \text{ ft}^2}{\text{min}}$

(C) Decreases by $\dfrac{240 \text{ ft}^2}{\text{min}}$

(D) Increases by $\dfrac{7.3 \text{ ft}^2}{\text{min}}$

8. $\displaystyle\int \frac{\left(\sqrt{x}-x\right)^2}{\sqrt{x}}\,dx =$

(A) $\dfrac{\left(\sqrt{x}-x\right)^3}{3}+C$

(B) $\dfrac{2}{3}x^{3/2}-x^2+\dfrac{2}{5}x^{5/2}+C$

(C) $\dfrac{2}{3}x^{3/2}-x^2+\dfrac{3x^{1/2}}{2}+C$

(D) $\dfrac{2}{3}x^{3/2}-x^2-\dfrac{1}{x^{3/2}}+C$

9. The graph of the even function $f(x)$, comprised of straight lines, is given below. Which of the following is true?

I. $\displaystyle\int_{-6}^{-1} f(x)\,dx = \int_{1}^{6} f(x)\,dx$

II. $\displaystyle\int_{-2}^{-1} f(x)\,dx = \int_{1}^{0} f(x)\,dx$

III. $\displaystyle\int_{0}^{1} f(x)\,dx - \int_{1}^{6} f(x)\,dx = 2\int_{-4}^{-2} |f(x)|\,dx$

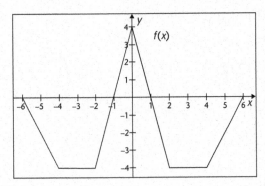

(A) I and II only

(B) I and III only

(C) II and III only

(D) I, II, and III

10. $\displaystyle\int_{\sqrt{e}}^{e} \frac{1}{t\ln t}\,dt =$

(A) $\dfrac{1}{e} - \dfrac{1}{\sqrt{e}}$

(B) $\dfrac{1}{2}$

(C) $\ln 2$

(D) $\ln\dfrac{1}{2}$

11. A particle travels on a straight line with velocity $v(t)$ as given in the figure to the right. In what interval(s) is the particle speeding up?

I. $(0, p)$

II. (p, q)

III. (q, r)

(A) III only

(B) II and III only

(C) I and III only

(D) II only

12. If $f(x) = \dfrac{x-1}{x^3+1}$, find $\displaystyle\lim_{h\to 0}\dfrac{f(1+h)-f(1)}{h}$

(A) $\dfrac{1}{2}$

(B) -1

(C) $\dfrac{1}{3}$

(D) does not exist

13. If $f(x) = x(\sin 2x - \cos 3x)$, how many locations on $[0, \pi]$ is the mean-value theorem satisfied? (Graphing calculator permitted.)

(A) 0

(B) 1

(C) 2

(D) 3

14. In Florida, there is a threat of a hurricane and people go to a large market for bottled water. Throughout a five-hour time period, 300 cases of water per hour are sold. The rate that the shelves are restocked is given by $R(t)$ as shown in the figure below. If there are 200 cases of water on the shelves at the start of the time period, what is the minimum number of cases on the shelves in the five-hour time period?

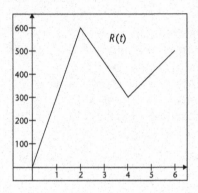

(A) 0

(B) 50

(C) 150

(D) 350

15. Find $\dfrac{d^2y}{dx^2}$ at (3, 2) for $2x^2 + 3y^2 = 30$.

(A) −1

(B) $\dfrac{-5}{6}$

(C) $\dfrac{1}{6}$

(D) $\dfrac{-2}{3}$

16. If $f(x) = e^{2x}(2x^2 - 6x + 5)$, for what values of x is f increasing and concave down?

    (A) $(-\infty, 0)$ only

    (B) $(0, 1)$

    (C) $(-\infty, 0), (1, \infty)$

    (D) $(0, \infty)$

17. The differentiable curve $f(x)$ is shown on the figure at right with 3 roots on the given interval I. At how many locations on the interval I does the graph of $|f(x)|$ fail to be differentiable?

    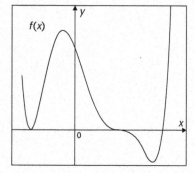

    (A) 0

    (B) 1

    (C) 2

    (D) 3

18. The difference in maximum acceleration and minimum acceleration attained on the interval $0 \le t \le 9$ by the particle whose velocity is given by

    $v(t) = 2t^{5/2} - 40t^{3/2} + 60t - 9$ is

    (A) 80

    (B) 45

    (C) 35

    (D) 4

19. The area bounded by $f(x) = 2\sqrt{x}$, $x = 0$, and $y = 6$ is rotated about the line $y = 6$. Find the volume of the solid.

    (A) 18π

    (B) 54π

    (C) 108π

    (D) 162π

20. The slope field below could be based on what differential equation?

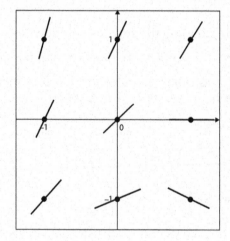

    (A) $\dfrac{dy}{dx} = ye^{-x}$

    (B) $\dfrac{dy}{dx} = e^y + x$

    (C) $\dfrac{dy}{dx} = e^y - x$

    (D) $\dfrac{dy}{dx} = e^{xy} - x$

21. Which of the following improper integrals are convergent?

    I. $\displaystyle\int_0^\infty xe^{-x}\,dx$

    II. $\displaystyle\int_2^\infty \frac{1}{x^2-x}\,dx$

    (A) I only

    (B) II only

    (C) Both I and II

    (D) Neither

22. Let $y = f(x)$ be the solution to the differential equation $\dfrac{dy}{dx} = x^2 - y^2 + 2$ with initial condition $f(-1) = -2$. What is the approximation for $f(0)$ if Euler's method is used, starting at $x = -1$ with step size of 0.5?

    (A) –4.5

    (B) –2.5

    (C) –0.25

    (D) –18.25

23. Let f be a function having derivatives for all orders of real numbers. The first three derivatives of f at $x = 0$ are given in the table below. If the third-degree Taylor polynomial centered at $x = 0$ approximates $f\left(-\dfrac{1}{2}\right) \approx -2$, what is the value of k?

x	$f(x)$	$f'(x)$	$f''(x)$	$f'''(x)$
0	–5	3	4	k

    (A) –384

    (B) –28

    (C) 96

    (D) –192

24. Find the behavior of the curve at $t = \dfrac{\pi}{4}$ of the parametric equations $x = 3\sin t$ and $y = \cos t$.

    (A) Increasing, concave up

    (B) Increasing, concave down

    (C) Decreasing, concave up

    (D) Decreasing, concave down

25. If convergent, find the interval of convergence for the series $1 - \dfrac{2(x+1)}{5} + \dfrac{3(x+1)^2}{25} - \dfrac{4(x+1)^3}{125} + \dots.$

    (A) (–6, 4)

    (B) [–6, 4]

    (C) [–5, 5]

    (D) divergent

ANSWERS

1. (C) is the correct answer.

    Average rate of change of f: $\dfrac{4^3 - (-2)^3}{4 - (-2)} = \dfrac{64 + 8}{6} = 12$

    Average value of f: $\dfrac{\displaystyle\int_{-2}^{4} x^3\, dx}{4 - (-2)} = \dfrac{\left[\dfrac{x^4}{4}\right]_{-2}^{4}}{6} = \dfrac{64 - 4}{6} = 10$

    $12 - 10 = 2$

    (A) incorrectly states that the average value and average rate of change are the same

    (B) calculates the integral with numerator 64 – 8

    (D) neglects the denominator of 6 in both fractions

2. (A) is the correct answer.

    $g(4) = \sqrt{4}\,f(4) = 2(8) = 16$

    Product rule: $g'(x) = \sqrt{x}f'(x) + \dfrac{f(x)}{2\sqrt{x}} \Rightarrow$

    $$g'(4) = \sqrt{4}f'(4) + \dfrac{f(4)}{2\sqrt{4}} = 2(-3) + \dfrac{8}{2(2)} = -4$$

    Tangent line: $y - 16 = -4(x - 4) \Rightarrow y - 16 = -4x + 16 \Rightarrow$
    $\qquad\qquad y = -4x + 32 \Rightarrow g(4.5) \approx -4(4.5) + 32 = 14$

    (B) only gives first half of product rule

    (C) incorrectly takes derivative of g as $\dfrac{f(x)}{2\sqrt{x}}$

    (D) incorrectly cancels the 16s

3. (B) is the correct answer.

    $$g'(x) = \frac{d}{dx}\int_0^x f(t)\,dt = f(x).$$

    $g'(x): \text{---------}0\,\text{+++++}\,0\,\text{---}$
    $\qquad\quad -6 \qquad\quad -2 \qquad\quad 5 \quad 6$

    The absolute minimum occurs at $x = -2$ or $x = 6$. Since the area under f between 0 and 6 is mostly positive and $\displaystyle\int_0^{-2} f(t)\,dt$ is negative, the absolute minimum occurs at $x = -2$.

    (A) $x = -3$ is one location where f has its absolute minimum, not g

    (C) $x = 5$ is where g has its absolute maximum

    (D) $x = 6$ was a candidate for the location of the absolute minimum

4. (C) is the correct answer.

$$\lim_{x \to 1} \frac{x \sin \pi x - \sin \pi x - x + 1}{x \cos \pi x - \cos \pi x - x + 1} = \frac{\sin \pi - \sin \pi - 1 + 1}{\cos \pi - \cos \pi - 1 + 1} =$$

$$\frac{0 - 0 + 0}{-1 + 1 + 0} = \frac{0}{0}$$

This could lead to a conclusion of answer (D) but there are other things to try.

Factorization: $\lim_{x \to 1} \dfrac{\sin \pi x (x - 1) - (x - 1)}{\cos \pi x (x - 1) - (x - 1)} =$

$$\lim_{x \to 1} \frac{(x - 1)(\sin \pi x - 1)}{(x - 1)(\cos \pi x - 1)}$$

$$\lim_{x \to 1} \frac{(\sin \pi x - 1)}{(\cos \pi x - 1)} = \frac{\sin \pi - 1}{\cos \pi - 1} = \frac{0 - 1}{-1 - 1} = \frac{1}{2}$$

L'Hôspital's Rule: $\lim_{x \to 1} \dfrac{\pi x \cos \pi x + \sin \pi x - \pi x \cos \pi x - 1}{-\pi x \sin \pi x + \cos \pi x + \pi x \sin \pi x - 1} =$

$$\lim_{x \to 1} \frac{\sin \pi x - 1}{\cos \pi x - 1} = \frac{0 - 1}{-1 - 1} = \frac{1}{2}$$

5. (D) is the correct answer.

LRS: $3(-6 - 1 + 2 + 5) = 0$ $|30 - 0| = 30$

RRS: $3(-1 + 2 + 5 + 7) = 39$ $|30 - 39| = 9$

Trap: $\dfrac{3}{2}(-6 - 2 + 4 + 10 + 7) = \dfrac{39}{2}$ $\left|30 - \dfrac{39}{2}\right| = 10.5$

MP: $6(-1 + 5) = 24$ $|30 - 24| = 6$

6. (D) is the correct answer.

$f'(x) = 3(3 - 2\sin x^2)^2 \, (-2\cos x^2)(2x)$

$-12x(3 - 2\sin x^2)^2 \, (\cos x^2)$

Answers (A), (B), and (C) neglect some aspect of the chain rule.

7. (A) is the correct answer.

Since a and b are constants, $A = 8(6)\sin\theta = 48\sin\theta$

$$\frac{dA}{dt} = 48\cos\theta\frac{d\theta}{dt} = 48\cos120°\left(\frac{10°}{\min}\right)\left(\frac{\pi}{180°}\right)$$

$$\frac{dA}{dt} = 48\left(\frac{-1}{2}\right)\left(\frac{\pi}{18}\right) = \frac{-4\pi}{3} \approx -4.2$$

(B) does not incorporate $\dfrac{d\theta}{dt}$

(C) does not change $\dfrac{d\theta}{dt}$ to radians

(D) uses $\sin\theta$ instead of $\cos\theta$ in the derivative

8. (B) is the correct answer.

This cannot be done by u-substitution as if

$u = x^{1/2} - x, du = \dfrac{1}{2x^{1/2}} - 1$

To integrate, the numerator must first be expanded

$$\int\frac{x - 2x^{3/2} + x^2}{x^{1/2}}\,dx = \int\left(x^{1/2} - 2x + x^{3/2}\right)dx =$$

$$\frac{2}{3}x^{3/2} - x^2 + \frac{2}{5}x^{5/2} + C$$

(A) incorrectly uses u-substitution

(C) and (D) incorrectly use the power rule on the last term

9. (D) is the correct answer. An even function means it is symmetric to the y-axis.

   I.   Both represent identical trapezoids below the axis.

   II. $\displaystyle\int_{-2}^{-1} f(x)\,dx$ is the negative area of the triangle.

Since $\displaystyle\int_{0}^{1} f(x)\,dx$ represents the positive area of

the triangle, $\displaystyle\int_{1}^{0} f(x)\,dx$ is the negative area.

III. $\int\limits_{0}^{1} f(x)\,dx = 2$ while $\int\limits_{1}^{6} f(x)\,dx = -\frac{1}{2}(4)(5+2) = -14$

and $2 - (-14) = 16$

$2\int\limits_{-4}^{-2}|f(x)|\,dx = 2(8) = 16$

10. (C) is the correct answer.

$$\int\limits_{\sqrt{e}}^{e} \frac{1}{t\ln t}\,dt = \int\limits_{1/2}^{1} \frac{1}{u}\,du = \ln|u|\Big|_{1/2}^{1} \qquad u = \ln t \quad du = \frac{1}{t}\,dt$$

$$\ln 1 - \ln\frac{1}{2} = 0 - (\ln 1 - \ln 2) = \ln 2 \quad t = \sqrt{e}, u = \frac{1}{2} \quad t = e, u = 1$$

(A) is the result of not changing the limits

(B) integrates incorrectly

(D) incorrectly reverses order of limits

11. (C) is the correct answer.

Particle speeding up when velocity and acceleration have the same signs. Velocity negative on $(0, p)$ and (p, q) and positive on (q, r). Acceleration, which is the derivative (slope) of v, is negative on $(0, p)$ and positive on (p, q) and (q, r). $v(t)$ and $a(t)$ have same signs on $(0, p)$ and (q, r).

(A) is an interval where velocity is positive, which is not the same as speeding up.

(B) are intervals where acceleration is positive, which is not the same as speeding up.

12. (A) is the correct answer.

$\lim\limits_{h\to 0}\dfrac{f(1+h)-f(1)}{h}$ is the definition of the derivative of the function f at $x = 1$: $f'(1)$

Quotient rule: $f'(x) = \dfrac{(x^3+1)-(x-1)(3x^2)}{(x^3+1)^2} = \dfrac{-2x^3+3x^2+1}{(x^3+1)^2}$

$f'(1) = \dfrac{-2+3+1}{2^2} = \dfrac{1}{2}$

Choice (B) is the result of finding $f(0)$

Choice (C) is the result of not using the quotient rule, just taking derivatives of numerator and denominator

13. (D) is the correct answer.

The function is differentiable on $[0, \pi]$ so the MVT holds

$f'(x) = x(2\cos 2x + 3\sin 3x) + \sin 2x - \cos 3x$

$\dfrac{f(\pi)-f(0)}{\pi-0} = \dfrac{\pi(\sin 2\pi - \cos 3\pi)}{\pi} = \dfrac{\pi[0-(-1)]}{\pi} = 1$

$x(2\cos 2x + 3\sin 3x) + \sin 2x - \cos 3x = 1$

$x(2\cos 2x + 3\sin 3x) + \sin 2x - \cos 3x - 1 = 0$

Graphing this equation on $[0, \pi]$ has 3 roots

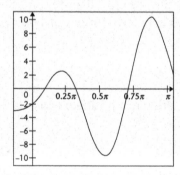

14. (B) is the correct answer.

To determine the minimum number of cases on the shelves, we must write an expression for the number of cases on shelves at any time t: $200 + \int_0^t [R(k) - 300] dk$.

Take the derivative to maximize/minimize this expression. This uses the Second Fundamental theorem.

$$\frac{d}{dt}\left[200 + \int_0^t [R(k) - 300] dk\right] = R(t) - 300 = 0 \Rightarrow$$

$R(t) = 300 \Rightarrow t = 1, t = 4$

$R(t) - 300: ------- 0 +++++++ 0 ++++++$

$$\begin{array}{cccc} 0 & 1 & 4 & 5 \end{array}$$

Since $R(t) - 300$ is negative if $t < 1$ and positive if $t > 1$, the minimum number of cases on the shelves occurs at $t = 1$. At $t = 1$, the number of cases that were added to the shelves is $\frac{1}{2}(300) = 150$.

200 were originally there and 300 were sold so there are $200 + 150 - 300 = 50$ cases on the shelves.

15. (B) is the correct answer.

Implicit differentiation: $4x + 6y\frac{dy}{dx} = 0 \Rightarrow \frac{dy}{dx} = \frac{-4x}{6y} = \frac{-2x}{3y}$

At (3, 2), $\frac{dy}{dx} = \frac{-6}{6} = -1$

$$\frac{d^2y}{dx^2} = \frac{3y(-2) + 2x\left(3\frac{dy}{dx}\right)}{9y^2}$$

At (3, 2), $\frac{d^2y}{dx^2} = \frac{6(-2) + 6(3)(-1)}{9(4)} = \frac{-12 - 18}{36} = \frac{-5}{6}$

(A) gives $\frac{dy}{dx}$, not $\frac{d^2y}{dx^2}$

(C) neglects using $\frac{dy}{dx}$ in formula for $\frac{d^2y}{dx^2}$

(D) does not use quotient rule in finding $\frac{d^2y}{dx^2}$

16. (B) is the correct answer.

To find when *f* is increasing and decreasing, make a sign chart for $f'(x)$

$$f'(x) = e^{2x}(4x-6)+(2x^2-6x+5)(2e^{2x})$$
$$= e^{2x}(4x-6+4x^2-12x+10)$$
$$4e^{2x}(x^2-2x+1) = 4e^{2x}(x-1)^2 : \underset{1}{++++++0+++++}$$

To find when *f* is concave up and down, make a sign chart for $f''(x)$

$$f''(x) = 4\left[e^{2x}(2x-2)+(x^2-2x+1)(2e^{2x})\right]$$
$$= 4e^{2x}(2x-2+2x^2-4x+2)$$
$$4e^{2x}(2x^2-2x) = 8xe^{2x}(x-1) : \underset{0 \qquad 1}{+++0-----0+++}$$

Function increasing if $x \neq 1$ and is concave down on (0, 1).

17. (B) is the correct answer.

The graph of $|f(x)|$ is below. Only the point farthest to the right on the *x*-axis is a cusp point and thus fails differentiability.

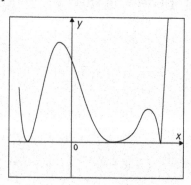

18. (A) is the correct answer.

    To find extrema for $a(t)$, we must set $a'(t) = 0$

    $$a(t) = 5t^{3/2} - 60t^{1/2} + 60$$

    $$a'(t) = \frac{15\sqrt{t}}{2} - \frac{30}{\sqrt{t}} = 0 \Rightarrow 15t - 60 = 0 \Rightarrow 15t = 60 \Rightarrow t = 4$$

    $a(0) = 60$
    $a(4) = 5(8) - 60(2) + 60 = -20$
    $a(9) = 5(27) - 60(3) + 60 = 15$
    Difference = $60 - (-20) = 80$

    (B) 45 is the difference between $a(0)$ and $a(9)$

    (C) 35 is the difference between $a(9)$ and $a(4)$

    (D) $t = 4$ is the difference in time between the absolute minimum and maximum

19. (B) is the correct answer.

    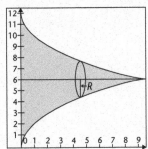

    $$2\sqrt{x} = 6 \Rightarrow \sqrt{x} = 3 \Rightarrow x = 9 \qquad \text{Disks: } R = 6 - 2\sqrt{x}$$

    $$V = \pi \int_0^9 \left(6 - 2\sqrt{x}\right)^2 dx = \pi \int_0^9 \left(36 - 24x^{1/2} + 4x\right) dx$$

    $$V = \pi \left[36x - 16x^{3/2} + 2x^2\right]_0^9 = \pi\left[324 - 16(27) + 162\right] = 54\pi$$

    (A) does not square integrand

    (C) incorrectly doubles the correct answer for area above $y = 6$

    (D) incorrectly treats the problem as washers:

    $$\pi \int_0^9 \left[6^2 - \left(2\sqrt{x}\right)^2\right] dx$$

20. (C) is the correct answer.

At (1, 0), $\frac{dy}{dx} = 0$, eliminating (B). At (0, 0), $\frac{dy}{dx} = 1$,

eliminating (A). At (0, –1), (D) $\frac{dy}{dx} = 1$, which is clearly incorrect.

21. (C) is the correct answer.

I. Integration by parts: $\int xe^{-x}\,dx = -xe^{-x} + \int e^{-x}\,dx$

$$\left[-xe^{-x} - e^{-x}\right]_0^\infty = \left[\frac{-x-1}{e^x}\right]_0^\infty \qquad u = x \qquad v = -e^{-x}$$
$$\qquad\qquad\qquad\qquad\qquad\qquad du = dx \qquad dv = e^{-x}dx$$

By L'Hôspital's rule: $0 - (-1) = 1$

II. Integration with partial fractions: $\int \frac{1}{x^2 - x}\,dx =$

$$\int \frac{1}{x(x-1)}\,dx = \int\left(\frac{-1}{x} + \frac{1}{x-1}\right)dx$$

$$\ln|x-1| - \ln|x| = \left[\ln\left|\frac{x-1}{x}\right|\right]_2^\infty$$

By L'Hôspital's rule: $\ln 1 - \ln\left(\frac{1}{2}\right) = 0 - (0 - \ln 2) = \ln 2$

22. (A) is the correct answer.

x	y	dy/dx	dy
–1	–2	–1	–0.5
–0.5	–2.5	–4	–2
0	–4.5		

(B) approximates $f(-2)$

(C) approximates $\frac{dy}{dx}$ at $x = -2$

(D) approximates $\frac{dy}{dx}$ at $x = 0$

23. (D) is the correct answer.

$$f(x) \approx P_3(x) = f(0) + f'(0)x + \frac{f''(0)}{2!}x^2 + \frac{f'''(0)}{3!}x^3$$

$$f\left(-\frac{1}{2}\right) \approx P_3\left(-\frac{1}{2}\right) = -5 + 3\left(-\frac{1}{2}\right) + \frac{4}{2!}\left(-\frac{1}{2}\right)^2 + \frac{k}{3!}\left(-\frac{1}{2}\right)^3 = -2$$

$$f\left(-\frac{1}{2}\right) \approx -5 - \frac{3}{2} + 2\left(\frac{1}{4}\right) + \left(\frac{k}{6}\right)\left(-\frac{1}{8}\right) = -5 - \frac{3}{2} + \frac{1}{2} - \frac{k}{48} = -2$$

$$-6 - \frac{k}{48} = -2 \Rightarrow \frac{k}{48} = -4 \Rightarrow k = -192$$

(A) makes an algebraic error

(B) ignores the factorials

(C) approximates $f\left(\dfrac{1}{2}\right)$

24. (D) is the correct answer.

$$\frac{dy}{dx} = \frac{dy/dt}{dx/dt} = \frac{-\sin t}{3\cos t} = -\frac{1}{3}\tan t$$

$$\frac{d^2y}{dx^2} = \frac{d(dy/dx)/dt}{dx/dt} = \frac{-1}{3}\left(\frac{\sec^2 t}{3\cos t}\right) = \frac{-1}{9\cos^3 t}$$

$$\frac{dy}{dx}\bigg|_{t=\pi/4} = \frac{-1}{3}\tan\frac{\pi}{4} = \frac{-1}{3}(1) = \frac{-1}{3} \quad \text{so curve is decreasing}$$

$$\frac{d^2y}{dx^2}\bigg|_{t=\pi/4} = \frac{-1}{9\left(\cos\frac{\pi}{4}\right)^3} = \frac{-1}{9\left(\frac{\sqrt{2}}{2}\right)} = \frac{-2}{9\sqrt{2}} \quad \text{so curve is concave down}$$

25. (A) is the correct answer.

To determine this, write the series as a formula:

$$\sum_{n=0}^{\infty}(-1)^n\frac{(n+1)(x+1)^n}{5^n}$$

Ratio test: $\lim\limits_{n\to\infty}\left|\frac{(-1)^{n+1}(n+2)(x+1)^{n+1}}{5^{n+1}}\cdot\frac{5^n}{(-1)^n(n+1)(x+1)^n}\right|$

$\lim\limits_{n\to\infty}\left|\frac{x+1}{5}\right|<1\Rightarrow|x+1|<5$ so $-6<x<4$

$x=-6:1+2+3+4+...$ which is divergent as the terms are getting larger.

$x=4:1-2+3-4+...$ which is alternating but the terms are not getting smaller so it is divergent

The interval of convergence is $(-6, 4)$.

The Free-Response Questions

You need to work out the free-response questions, step by step. These questions often involve graphs. Partial credit is given for steps of these exercises. It is extremely important that you show all of the steps. Doing so allows the AP® Reader to follow your logic and provides more opportunity for you to earn every point for which you are eligible. In fact, answers that appear without proper supporting logic usually earn no credit! Also, when explaining your process, use complete sentences.

Each part of every exercise has an indicated workspace. Show all work for that part in that space only! Be sure to write neatly, so the reader can follow your work. If you make an error, it is recommended that you cross it out rather than erase it. If you replace the crossed-out material with another answer, the reader will only look at that. But if you leave it blank, the reader might look at the crossed-out section and find something in there worth a point. So don't erase unless you are certain it is incorrect.

The free-response section will include 6 problems to be completed in 90 minutes. You will use pencil or dark blue or black ink to write out your work on these questions. Although the parts of each question are not necessarily equally weighted, each question as a whole is equally weighted and worth 9 points.

THE FREE-RESPONSE SECTIONS

The free-response section of the AP® Calculus examinations has two parts. A graphing calculator is allowed on Part A, which has two questions. You will have 30 minutes to work on Part A.

Part B has 4 questions and use of the graphing calculator is prohibited. You have 60 minutes to work on this part. During this hour, you may continue to work on Part A, without the use of the graphing calculator.

Action instructions are commonly used in the free-response questions. They give students clear direction as to what is expected in the response. Here are typical action instructions and examples of their use.

Action Instructions	Example
Approximate: Typically used when exact calculations are too difficult to find or impossible to find with the given information.	Given the accompanying table of x versus $f(x)$, approximate $f'(2)$.
Calculate: Find a numerical or algebraic answer to a question. Instructions can also be expressed as "how many?" or "find the value."	Calculate the volume when the area between $f(x)$ and $g(x)$ is rotated about the x-axis.
Determine: Apply an appropriate definition, theorem, or test to indicate values, intervals, or solutions where a solution might exist.	Given the power series, determine its interval of convergence.
Evaluate: Apply mathematical processes, including rounding procedures, to find the value of an expression at a given point or interval.	Given $g(x) = \int_0^x f(t)\,dt,$ evaluate $g(2)$.
Explain: Use appropriate definitions or theorems to provide rationales for solutions and conclusions. "Explain" instructions may also be phrased as "give a reason why."	Explain why $f(x)$ is differentiable within its domain.

Action Instructions	Example
Identify/Indicate: Provide information about a specific topic without elaboration or explanation.	Identify the time intervals when the particle is speeding up.
Indicate units of measure: If this direction is given, it is usually worth one point in the grading. So if you have no idea how to do the problem, just indicating its units of measurement can gain you a point.	Find how fast the rate of change of the speed of the glacier is decreasing. Indicate units of measure.
Interpret . . . In the context of the problem: Describe the connection between a mathematical expression or solution and its meaning within the realistic context of a problem, often including consideration of units.	If $r(t)$ represents the rate that people board a cruise ship, t measured in minutes, interpret $\int_0^{60} r(t)\,dt$ in context of the problem situation.
Interpret (when given a representation): Identify mathematical information represented graphically, symbolically, verbally, and/or numerically.	If the accompanying graph represents the velocity of a car as it attempts to move out of a tight parking spot, interpret the section of the curve when the graph is above the t-axis in the context of the problem.
Justify: Identify a logical sequence of mathematical definitions, theorems, or test, to support an argument or conclusion, explain why these apply, and then apply them. While you can use sign charts or graphs to help you solve a problem, justifications are always in sentences.	Given $f'(x)$, for what values of x is the graph of f concave up? Justify your answer.
Represent: Use appropriate graphs, symbols, words, and/or tables of numerical values to describe mathematical concepts, characteristics, and/or relationships.	If $f(x) = 1 + \dfrac{x}{2} + \dfrac{x^2}{6} + \dfrac{x^3}{24} + \ldots,$ represent $f(x)$ using a common function.

Action Instructions	Example
Show that: Rather than asking for a result, the result is given and students must show how it was derived. This occurs when the testers wish you to use a correct answer on another part of the question.	If $x^2 - 2y + y^2 = 8$, show that $\dfrac{dy}{dx} = \dfrac{x}{1-y}$.
Show the work that leads to your conclusion: Rather than just state an answer, show all the steps along the way. Since the word itself is being graded, don't skip steps along the way.	Use the ratio test to find the interval of convergence for the given power series. Show the work that leads to your conclusion.
Use: Solve a problem using a specific methodology.	Use separation of variables to find an expression for h in terms of t.
Verify: Confirm that the conditions of a mathematical definition, theorem or test are in order to explain why it applies in a given situation. Alternately confirm that solutions are accurate and appropriate.	Verify that the Mean-Value Theorem is appropriate for $f(x)$ in the interval $[-4, 4]$ and find the value of c guaranteed by the MVT.
Write, but do not evaluate: Create a mathematical expression that gives the solution to a problem without calculating its numerical value.	Write, but do not evaluate, an integral expression that determines the value of c in which the unbounded area under $f(x)$ to the right of the y-axis is divided into two equal areas.

EXAMPLE QUESTION FOUND IN SECTION II, PART A:

Let R be the region bounded by the graphs of $y = 10e^{-x}$, $y = \ln x$, $x = 1$, and $x = 2$.

    A. What is the area of the region bounded by the curves?

B. The region *R* is the base of a solid with cross sections perpendicular to the *x*-axis are rectangles whose height is 4 times its base. Set up, but do not evaluate, an integral expression in terms of a single variable representing the volume of this solid.

C. Set up, but do not solve, the integral expression in terms of a single variable for the volume of the solid generated by revolving this enclosed region around the line $y = -1$.

Solution:

A. What is the area of the region bounded by the curves?

Take a look at the graph.

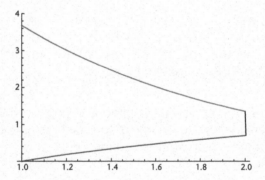

Knowing that the graph of $y = 10e^{-x}$ is above the graph of $y = \ln x$, you set up your definite integral as $\int_{1}^{2}(10e^{-x} - \ln x)dx$.

The smaller *x*-value is the lower limit of integration; the larger *x*-value is the upper limit of integration. The integrand is the top graph minus the bottom graph.

$$\int_{1}^{2}(10e^{-x} - \ln x)dx \approx 1.939$$

This part would be worth 3 points: 1 for the graph, 1 for setting up the integral, and 1 for finding the numerical solution.

B. The region *R* is the base of a solid with cross sections perpendicular to the *x*-axis are rectangles whose height is 4 times its base. Set up, but do not evaluate, an integral expression in terms of a single variable representing the volume of this solid.

The area of the rectangle is base • height. The base is $10e^{-x} - \ln x$. The height is 4 times the base or $4(10e^{-x} - \ln x)$.

Hence the area of the rectangle is $4(10e^{-x} - \ln x)^2$.

$4(10e^{-x} - \ln x)$

$A = 4(10e^{-x} - \ln x)^2$

$10e^{-x} - \ln x$

So $V = \int_{1}^{2} 4(10e^{-x} - \ln x)^2\, dx$ or $V = 4\int_{1}^{2} (10e^{-x} - \ln x)^2\, dx$. This part would be worth 3 points: 1 for the limits of integration, 1 for the constant, and 1 for the integrand.

C. Set up, but do not solve, the integral expression in terms of a single variable for the volume of the solid generated by revolving this enclosed region around the line $y = -1$.

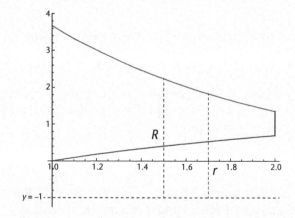

Finding the volume by revolving this region around the line

$y = -1$ requires the method of washers: $V = \pi\int_{a}^{b} [R^2 - r^2]\,dx$.

The R in this exercise is the distance between $y = -1$ and the curve $y = 10e^{-x}$. So, $R = 1 + y = 1 + 10e^{-x}$. The r in this exercise is the distance between $y = -1$ and the curve $y = \ln x$, making $r = 1 + y = 1 + \ln x$. Putting this all together, you get:

$$V = \pi \int_{1}^{2} [(1 + 10e^{-x})^2 - (1 + \ln x)^2] dx.$$

This part would be worth 3 points: 1 for the limits of integration, 1 for the constant, and 1 for the integrand. If the integrand is not in the form of $R^2 - r^2$, maximum of 1 point.

SECTION II, PART A: EXAMPLE II:

Let f be the function given by $f(x) = \dfrac{1}{2}x^3 - \dfrac{1}{2}x^2 - 1$.

A. Write an equation of the tangent line at $x = -1$.
B. List and identify all relative extreme points, both minimum and maximum.
C. What is the inflection point?

Solution:

A. Write an equation of the tangent line at $x = -1$.

The slope of the tangent line is the first derivative, $f'(x) = \dfrac{3}{2}x^2 - x$, evaluated at $x = -1$:

$f'(-1) = \dfrac{3}{2}(-1)^2 - (-1) = \dfrac{3}{2} + 1 = \dfrac{5}{2}$. The point when $x = -1$ is $(-1, -2)$. So the equation of the tangent line is given by: $y + 2 = \dfrac{5}{2}(x + 1)$ or $y = \dfrac{5}{2}x + \dfrac{1}{2}$. This part would be worth 2 points: 1 for the derivative and 1 for writing the equation of the tangent line.

B. List and identify all relative extrema, both minimum and maximum, points.

The first derivative set equal to 0 gives the critical numbers:

$$f'(x) = \frac{3}{2}x^2 - x = 0$$

$$x\left(\frac{3}{2}x - 1\right) = 0$$

$$x = 0 \quad \text{and} \quad x = \frac{2}{3}$$

The critical points are $(0, -1)$ and $\left(\frac{2}{3}, \frac{-29}{27}\right)$. The second derivative is $f''(x) = 3x - 1$. Evaluating the second derivative at $x = 0$, the result is negative, which indicates a relative maximum point. Evaluating the second derivative at $x = \frac{2}{3}$, the result is positive, which indicates a relative minimum point. This part would be worth 4 points: 1 point for each critical value of x, 1 point for the actual extreme points, and 1 point each for its classification and second derivative justification. Note that a first derivative test justification is acceptable as well.

C. What is the inflection point? Justify your answer.

Setting the second derivative equal to 0, $f''(x) = 3x - 1 = 0$, $x = \frac{1}{3}$. The inflection point is $\left(\frac{1}{3}, \frac{-28}{27}\right)$. Since $f''(x) < 0$ if $x < \frac{1}{3}$ and $f''(x) > 0$ if $x > \frac{1}{3}$, there is a change of concavity at $x = \frac{1}{3}$, and thus there is an inflection point there. This part would beworth 3 points: 1 point for the x-value, 1 point for the y-value, and 1 point for the justification.

EXAMPLE QUESTION FOUND IN SECTION II, PART B:

The acceleration of a particle is $a(t) = -32$ ft/sec^2, the initial velocity of the particle is 64 ft/sec, and the initial height of the particle is 40 ft.

    A. What is the formula for the velocity of the particle at any time t?

    B. What is the formula for the position of the particle at any time t?

    C. What is the maximum height this particle reaches?

A. Given $a(t) = -32$, you should find its antiderivative as the velocity of the particle. The antiderivative of $a(t) = -32$ is $v(t) = -32t + C$. You know that the initial velocity is 64; that is $v(0) = 64$. This means

$$v(t) = -32t + C$$
$$v(0) = -32(0) + C = 64$$
$$C = 64$$
$$v(t) = -32t + 64$$

This should earn you 2 points, 1 for the antidifferentiation and 1 for finding the constant.

B. You now know that $v(t) = -32t + 64$. You take the antiderivative of $v(t)$ to get the position of the particle, $s(t)$, at any time t. The antiderivative is $s(t) = -16t^2 + 64t + C$. You know that the initial position, $s(0) = 40$. Hence,

$$s(t) = -16t^2 + 64t + C$$
$$s(0) = -16(0)^2 + 64(0) + C = 40$$
$$C = 40$$
$$s(t) = -16t^2 + 64t + 40$$

This exercise should also earn you 2 points, 1 for the anti-differentiation and 1 for finding this constant.

C. The maximum height of this particle is reached when its velocity is zero. (The particle instantaneously stops so that it can make its way back down.) Setting the velocity equal to zero, you get

$$v(t) = -32t + 64 = 0$$
$$-32t = -64$$
$$t = 2$$

You now know that the particle reaches its maximum height when $t = 2$. You need to find the maximum height. Place $t = 2$ into the position function:

$$s(t) = -16t^2 + 64t + 40$$
$$s(2) = -16(2)^2 + 64(2) + 40 = -16(4) + 64(2) + 40$$
$$= -64 + 128 + 40 = 104$$

You will earn 2 points for this exercise: 1 point for finding the time that the particle reaches its maximum height and 1 point for finding that maximum height.

The previous example is worth a total of 6 points.

SECTION II, PART B: EXAMPLE II:

Use $x^2 - xy + 4y^2 = 15$ for A – C.

A. Show that $\dfrac{dy}{dx} = \dfrac{-2x + y}{-x + 8y}$.

B. Find the point(s) where there are horizontal tangents to the curve. Show how you arrived at your answer.

C. What is $\dfrac{d^2y}{dx^2}$ at $x = 1$? Show how you arrived at your answer.

Solution:

A.
$$x^2 - xy + 4y^2 = 15$$
$$2x - 1y + \frac{dy}{dx}(-x) + 8y\frac{dy}{dx} = 0$$
$$-x\frac{dy}{dx} + 8y\frac{dy}{dx} = -2x + y$$
$$\frac{dy}{dx}(-x + 8y) = -2x + y$$
$$\frac{dy}{dx} = \frac{-2x + y}{-x + 8y}$$

This would earn 2 points, 1 for the differentiation and 1 for the simplification.

B. $\dfrac{dy}{dx} = \dfrac{-2x + y}{-x + 8y} = 0$ so $-2x + y = 0$ and $y = 2x$

Original equation: $x^2 - x(2x) + 4(2x)^2 = 15$
$15x^2 = 15$
$x = \pm 1$
$x = 1, y = 2 \quad x = -1, y = -2$
1 point for $y = 2x$, 1 point for using the original equation
1 point each for (1, 2) and (−1, −2)

C. $\dfrac{d^2y}{dx^2} = \dfrac{(-x+8y)\left(-2+\dfrac{dy}{dx}\right) - (-2x + y)\left(-1 + 8\dfrac{dy}{dx}\right)}{(-x+8y)^2}$

$\dfrac{d^2y}{dx^2}_{(1,2)} = \dfrac{(-1+16)(-2+0) - (-2+2)(-1+8(0))}{(-1+16)^2}$

$\dfrac{d^2y}{dx^2}_{(1,2)} = \dfrac{15(-2)}{15^2} = \dfrac{-2}{15}$

1 point for formula for $\dfrac{d^2y}{dx^2}$
1 point for $\dfrac{dy}{dx} = 0$
1 point for answer

SECTION II, PART B: EXAMPLE III:

The continuous function g is defined as $g(x) = \dfrac{e^{-\sin x} - 1}{\sin x}$ for $x \neq 0$ and $g(0) = -1$. The function g has derivatives of all orders at $x = 0$.

A. Write the first 4 terms and the general term for the Maclaurin series for $f(x) = e^{-x}$.

B. Use the Maclaurin series found in part A to write the first four nonzero terms and the general term for g about $x = 0$.

C. Find the radius of convergence for the series found in part B. Show your work that leads you to your conclusion.

D. If the first 4 terms of g were used to estimate the value of $g(k)$, show that the greatest possible error that would be generated for any value of k is less than $\dfrac{1}{100}$. Explain your reasoning.

Solutions:

A. $e^{-x} = 1 - x + \dfrac{x^2}{2!} - \dfrac{x^3}{3!} + \ldots + \dfrac{(-1)^n x^n}{n!} + \ldots$

This is a 2-point question: 1 point for the first 4 terms and 1 point for the general term.

B. $e^{-\sin x} = 1 - \sin x + \dfrac{(\sin x)^2}{2!} - \dfrac{(\sin x)^3}{3!} + \ldots + \dfrac{(-1)^n (\sin x)^n}{n!} + \ldots$

$e^{-\sin x} - 1 = -\sin x + \dfrac{(\sin x)^2}{2!} - \dfrac{(\sin x)^3}{3!} + \ldots + \dfrac{(-1)^{n+1} (\sin x)^{n+1}}{(n+1)!} + \ldots$

$g(x) = -1 + \dfrac{\sin x}{2!} - \dfrac{(\sin x)^2}{3!} + \dfrac{(\sin x)^3}{4!} + \ldots + \dfrac{(-1)^{n+1} (\sin x)^n}{(n+1)!} + \ldots$

This is worth 2 points, 1 for the first 4 terms and 1 for the general term.

C. $\lim\limits_{n\to\infty}\left|\dfrac{(-1)^{n+2}(\sin x)^{n+1}}{(n+2)!}\cdot\dfrac{(n+1)!}{(-1)^{n+1}(\sin x)^{n}}\right|=\lim\limits_{n\to\infty}\left|\dfrac{\sin x}{n+2}\right|=0$

So the radius of convergence of g is $(-\infty, \infty)$.

This question would be worth 3 points: 1 for setting up the ratio test, 1 for showing that the limit is zero, and 1 for the correct answer.

D. Since g is an alternating series that has the absolute value of its terms decreasing for any value of x, the error in estimating g using 4 terms is the 5th term. For any value of k, the largest value of $\sin k = 1$. So the largest possible error is $\dfrac{1}{5!} < \dfrac{1}{100}$.

This is worth 2 points. One point for the largest error is the $n + 1$st term and 1 for the answer with reasoning.

AP® CALCULUS SCORING

On the multiple-choice portion of the examination, total scores will be based upon the number of questions answered correctly. (There will be no deductions for incorrect answers and no points will be given for unanswered questions.) Before taking any examination, keep in mind how the test is scored. Since no points are deducted for incorrect answers, you should answer every question, even if you have to guess.

The scores for the multiple-choice and the free-response sections are equally weighted. Your score will be a combined score of the computer-scored multiple-choice portion and the AP® Reader-scored free-response portion. The highest possible score on the actual AP® examination is 5. Each college or university determines the necessary score for credit in a college-level course.

As is common on standardized examinations, students are not expected to be able to answer every question.

If you take the AP® Calculus BC examination, you will receive an AP® Calculus AB examination subscore since approximately 60% of the Calculus BC examination is at the Calculus AB level.

CALCULATOR FOR AP® CALCULUS

A calculator is allowed on certain sections of the AP® Calculus exams. You should be prepared to use a graphing calculator that has the following capabilities: plotting the graph of a function, finding the zeros of functions, and numerically calculate derivatives and definite integrals. If you have questions about the appropriateness of your calculator for use on the AP® Calculus examination you're taking, visit the AP® Calculus AB & BC home page at *www.collegeboard.org*.

Suppose you are asked to find the area between the graph of $y = 3 + 2x - x^2$ and the axis. Whether or not this problem appears in the calculator or non-calculator section of the exam makes a difference as to the approach you will take and what you will show.

However, no matter what, you will need to determine the function's x-intercepts. If you are permitted to use the calculator, you can simply graph the function as shown below to determine its x-intercepts. Then use the INTEGRAL command on the CALC menu to show the shading, or use the FNINT command from the main screen to evaluate the integral. None of this needs be shown. The only statement required to get full credit is $\int_{-1}^{3}(3 + 2x - x^2)dx = 10.667$ or $\int_{-1}^{3}(3 + 2x - x^2)dx = \dfrac{32}{3}$. You will be graded on the limits of integration, the integrand, and the mathematical answer.

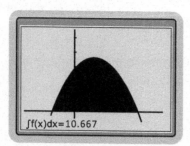

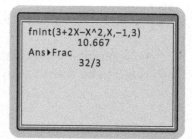

However, if this problem appears on the non-calculator section of the exam, you are required to show the work in finding the roots by setting $3 + 2x - x^2$ equal to zero. You then set up the area integral as $A = \int_{-1}^{3}\left(3 + 2x - x^2\right)dx$ and show necessary steps using the Fundamental Theorem to calculate it.

It is possible that this problem will appear on the calculator section but with the direction to *show the work that supports your answer*. In that case, the integration steps are important, as you will be graded on their presence and accuracy. You could use the calculator to verify your answer, but that is all. So it is very important to read the action instructions as discussed on pages 200–202 and pay attention to what you are required to do.

In addition, if you are required to provide a graph in the exam, regardless of whether the problem is in the calculator or non-calculator section, be sure the graph is labeled correctly and that its scale is clear. Calculator screens do not actually show a scale, so you must provide it or you will be docked a point.

What about approximating answers? You should have approximations correct to three decimal places, unless specified differently in an exercise. Do not round until the very end of the exercise!

Do not use "Calc-u-Speak" when answering questions. That is, do not use calculator terminology or explain what buttons you pressed. Statements like "Here is the graph in a ZOOM-6 window" mean nothing to the graders. The calculator should be used to support your work, but bear in mind that you are graded on your calculus knowledge, not on how you use the calculator.

—— SOLUTIONS FOR PRACTICE PROBLEMS ——

CHAPTER 2
SOLUTIONS

1. $y = \dfrac{1}{2}(x+1)^3 - 3$

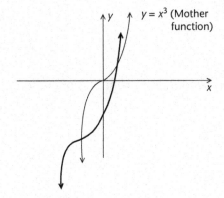

The mother function is widened, moved down 3 units and to the left one unit.

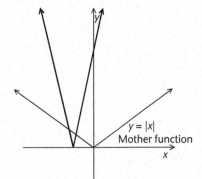

2. $y = 2|3x + 4|$

The mother function is narrowed and moved left $\dfrac{4}{3}$ units since $3x + 4 = 3\left(x + \dfrac{4}{3}\right)$.

3. $y = \sqrt{x - 6} + 1$

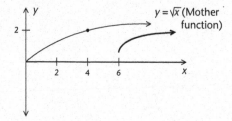

The mother function is moved 6 units right and one unit up.

4. $y = -\dfrac{3}{x} + 2$

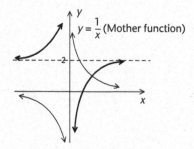

Mother function is reflected across x-axis, stretched by a factor of 3 and then moved up 2 units.

5. $y = e^{x+2} - 1$

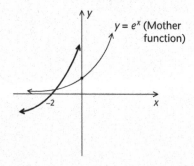

Mother function is moved 2 units left and one unit down.

6. $y = \ln(4 - x)$

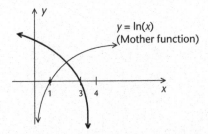

Mother function is reflected across *y*-axis and moved 4 units to the right.

CHAPTER 3
SOLUTIONS

1. (D) $-\dfrac{1}{8}$

   $\displaystyle\lim_{x \to \infty} \dfrac{1 - 3x + 6x^2 - x^{10}}{2 + 4x^4 - 8x^7 + 8x^{10}} = \lim_{x \to \infty} -\dfrac{x^{10}}{8x^{10}} = -\dfrac{1}{8}$. Use ratio of leading terms.

2. (A) 0

   $\displaystyle\lim_{x \to \infty} \dfrac{\sin x}{x} = 0$

   As $x \to \infty$, the denominator increases without bound whereas the numerator oscillates between –1 and 1.

3. (D) $-\infty$

   $\displaystyle\lim_{x \to 0^+} (\ln x) = -\infty$

   Use the graph of $y = \ln x$.

4. (C) $\dfrac{1}{6}$

$$\lim_{x \to 3}\dfrac{\sqrt{x+6}-3}{x-3}=\dfrac{1}{6}$$

Recognize the limit as the derivative of $f(x) = \sqrt{x+6}$ at $x = 3$.

So $f'(x) = \dfrac{1}{2\sqrt{x+6}}$ and $f'(3) = \dfrac{1}{2\sqrt{9}} = \dfrac{1}{6}$. or, without derivatives:

$$\lim_{x \to 3}\left(\dfrac{\sqrt{x+6}-3}{x-3}\right)\left(\dfrac{\sqrt{x+6}+3}{\sqrt{x+6}+3}\right)=\lim_{x \to 3}\dfrac{x-3}{(x-3)(\sqrt{x+6}+3)}=\lim_{x \to 3}\dfrac{1}{\sqrt{x+6}+3}$$

$$=\dfrac{1}{3+3}=\dfrac{1}{6}.$$

5. (D) Does not exist

$$\lim_{x \to 1}\dfrac{3}{x-1}=dne$$

From the graph of $y = \dfrac{3}{x-1}$ we see that $\lim_{x \to 1^-}\dfrac{3}{x-1}=-\infty$ and

$$\lim_{x \to 1^+}\dfrac{3}{x-1}=+\infty.$$

6. Vertical asymptote: $x = -3$; horizontal asymptote: $y = 3$

$f(x) = \dfrac{3x^2-9x}{x^2-9}$ $f(x) = \dfrac{3x(x-3)}{(x+3)(x-3)} = \dfrac{3x}{x+3}$ vertical

asymptote occurs when $x + 3 = 0$ or at $x = -3$.

$$\lim_{x \to \pm\infty}\dfrac{3x^2-9x}{x^2-9}=3 \text{ so horizontal asymptote at } y = 3.$$

7. Vertical asymptote: $x = \sqrt[3]{4}$; horizontal asymptote: $y = -1$

$f(x) = \dfrac{x^3+3x^2-1}{4-x^3}$ vertical asymptote occurs when $4 - x^3 = 0$

or at $x = \sqrt[3]{4}$ since $x = \sqrt[3]{4}$ is not a root of the numerator.

$\lim_{x \to \pm\infty} f(x) = -1$ so horizontal asymptote at $y = -1$.

8. (A) $y = 7$ is a horizontal asymptote of $f(x)$. The curve could have vertical asymptotes.

   (B) As $x \to -\infty$ $f(x)$ does not approach a horizontal asymptote, it decreases without bound. The curve could have vertical asymptotes.

   (C) $x = 4$ is a vertical asymptote of $f(x)$. The curve could have horizontal asymptotes.

9. $x < 1$

   $f(x) = \dfrac{2}{\sqrt{1-x}}$ is continuous when $1 - x > 0$ or $x < 1$.

10. Removable discontinuity at $x = -2$; nonremovable discontinuity at $x = 2$.

   $f(x) = \dfrac{x^2 + x + 6}{x^2 - 4} = \dfrac{(x+2)(x+3)}{(x+2)(x-2)}$ $\therefore$ $f(x)$ has a removable discontinuity at $x = -2$ and a nonremovable discontinuity at $x = 2$.

11. $x = -1,\ x = 0$

   Look at the graph of $f(x) = \begin{cases} 2 - x, & x < -1 \\ \dfrac{1}{x}, & -1 \le x \le 2 \\ \dfrac{1}{2}, & x > 2 \end{cases}$

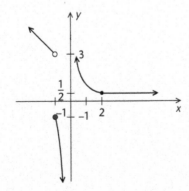

   The function is discontinuous at $x = 0$ and $x = -1$.

CHAPTER 4
SOLUTIONS

1. $y' = \dfrac{23}{(1-5x)^2}$

   $y' = \dfrac{(1-5x)(3) - (3x+4)(-5)}{(1-5x)^2} = \dfrac{3 - 15x + 15x + 20}{(1-5x)^2} = \dfrac{23}{(1-5x)^2}.$

2. $y' = \dfrac{5}{2\sqrt{5x+3}}$

   $\dfrac{d}{dx}(\ln(e^{\sqrt{5x+3}})) = \dfrac{d}{dx}(\sqrt{5x+3}) = \dfrac{5}{2\sqrt{5x+3}}.$

3. y' at $x = -1$ is $\dfrac{1}{3}$

   $3x - x^2y = 5y \rightarrow 3 - x^2y' - 2xy = 5y' \rightarrow 3 - 2xy = 5y' + x^2y' \rightarrow$

   $3 - 2xy = y'(5 + x^2) \rightarrow y' = \dfrac{3 - 2xy}{5 + x^2}.$ To evaluate y' at $x = -1$, we

   must substitute $x = -1$ into the original equation to find y.

   $3(-1) - (-1)^2 y = 5y \rightarrow -3 - y = 5y \rightarrow -3 = 6y \rightarrow y = -\dfrac{1}{2}.$

   So $y'\Big|_{\left(-1, -\frac{1}{2}\right)} = \dfrac{3 - 2(-1)\left(-\dfrac{1}{2}\right)}{5 + (-1)^2} = \dfrac{2}{6} = \dfrac{1}{3}.$

4. Derivative does not exist.

$y = x^2 - 4x \xrightarrow{\text{INVERSE}} x = y^2 - 4y \xrightarrow{\text{DERIVATIVE}} 1 = 2yy' - 4y' \rightarrow$

$1 = y'(2y - 4)$. Since $x = 2$ is on the original function, $y = 2$ is on

the inverse. So $y' = \dfrac{1}{2y - 4} \rightarrow y' = \dfrac{1}{2(2) - 4} = \dfrac{1}{0}$ dne. Graphically

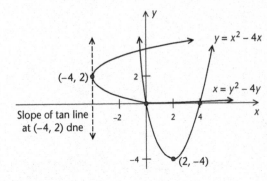

5. $h'(x) = f(x)g'(x) + g(x)f'(x)$

$h'(4) = f(4)g'(4) + g(4)f'(4)$

$h'(4) = -8(4) + 3\pi(3) = 9\pi - 32$

6. $f'(x) = 5\left(\dfrac{2x-1}{2x+1}\right)^4 \left[\dfrac{(2x+1)2 - (2x-1)2}{(2x+1)^2}\right]$

$f'(x) = 5\left(\dfrac{2x-1}{2x+1}\right)^4 \left[\dfrac{4}{(2x+1)^2}\right] = \dfrac{20(2x-1)^4}{(2x+1)^6}$

7. $h'(x) = f'(g(x))(g'(x))$

$h'(2) = f'(g(2))(g'(2))$

$h'(2) = f'(3)(-2) = 4(-2) = -8$

8. $y = \left[\cos(x^2)\right]^{1/2}$

$y' = \dfrac{1}{2}\left[\cos(x^2)\right]^{-1/2}\left[-\sin(x^2)\right](2x) = \dfrac{-x\sin(x^2)}{\sqrt{\cos x^2}}$

9. $2x - \dfrac{3}{y}\dfrac{dy}{dx} + 2y\dfrac{dy}{dx} = 0 \Rightarrow 2xy - 3\dfrac{dy}{dx} + 2y^2\dfrac{dy}{dx} = 0$

$(3 - 2y^2)\dfrac{dy}{dx} = 2xy \Rightarrow \dfrac{dy}{dx} = \dfrac{2xy}{3 - 2y^2}$

10. $y' = 3(e^x - 2x - 1)^2 (e^x - 2)$

CHAPTER 5
SOLUTIONS

1. The top of the ladder is slipping at a rate of .013 ft/sec.

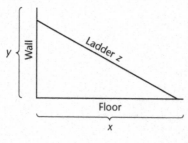

Floor

x

Let x = distance from the foot of the ladder to wall in feet

y = distance from top of the ladder to floor in feet.

Given: $z = 13$ ft, $\dfrac{dx}{dt} = 2$ in/sec,

$\dfrac{dy}{dt} = ?$ when $x = 1$ ft

$x^2 + y^2 = z^2 \rightarrow 2x\dfrac{dx}{dt} + 2y\dfrac{dy}{dt} = 0 \rightarrow 2(1)\left(\dfrac{1}{6}\right) + 2(\sqrt{168})\dfrac{dy}{dt}$

$= 0 \rightarrow \dfrac{dy}{dt} = -\dfrac{1}{3 \cdot 2\sqrt{168}} = -.013$ ft/sec

Notice that $\dfrac{dx}{dt} = 2$ in/sec was converted to $\dfrac{dx}{dt} = \dfrac{1}{6}$ ft/sec so that all units are in feet. Also, when $x = 12$ in $= 1$ ft, using the Pythagorean Theorem $y = \sqrt{168}$ ft. The answer is negative as the top of the ladder is getting closer to the ground.

2. $\dfrac{dp}{dt} = \dfrac{3}{160}$

$p\dfrac{dx}{dt} + x\dfrac{dp}{dt} - 10\dfrac{dp}{dt} = 3\dfrac{dx}{dt}$

$\dfrac{13}{4}(-3) + 50\left(\dfrac{dp}{dt}\right) - 10\left(\dfrac{dp}{dt}\right) = 3(-3) \rightarrow \dfrac{dp}{dt} = \dfrac{3 \text{ dollars}}{160 \text{ day}} \approx \dfrac{1.9 \text{ cents}}{\text{day}}$

3. $f'(x) = \dfrac{7}{(2x+1)^2} \Rightarrow f'(3) = \dfrac{1}{7}$ $f(3) = 1$

Tangent line: $y - 1 = \dfrac{1}{7}(x-3) \Rightarrow y = \dfrac{x+4}{7}$

$f(2.9) \approx y(2.9) = \dfrac{6.9}{7} = \dfrac{69}{70}$

4. Tangent line: $y + 3 = 5(x+1) \Rightarrow y = 5x + 2$

$f(-0.99) = 5(-0.99) + 2 = -2.95$

$f(-1.01) = 5(-1.01) + 2 = -3.05$

5. $\lim\limits_{x \to 0}\left(\dfrac{x\cos\pi x + \sin x}{x}\right) = \dfrac{0}{0}$

$\lim\limits_{x \to 0}\left(\dfrac{-\pi x \sin \pi x + \cos \pi x + \cos x}{1}\right) = 2$

6. $\lim\limits_{x \to \infty}\left(\dfrac{4x^2 - 5x + 2}{e^{5x} + \ln x}\right) = \dfrac{\infty}{\infty}$

$\lim\limits_{x \to \infty}\left(\dfrac{8x - 5}{5e^{5x} + \frac{1}{x}}\right) = \dfrac{\infty}{\infty} \Rightarrow \lim\limits_{x \to \infty}\left(\dfrac{8}{25e^{5x} - \frac{1}{x^2}}\right) = 0$

CHAPTER 6
SOLUTIONS

1. $c = \sqrt{3} + 1$

$$f'(c) = \frac{f(b) - f(a)}{b - a} \rightarrow -\frac{1}{(c-1)^2} = \frac{f(4) - f(2)}{4 - 2} \rightarrow$$

$$-\frac{1}{(c-1)^2} = \frac{-\frac{2}{3}}{2} \rightarrow -\frac{1}{(c-1)^2} = -\frac{1}{3} \rightarrow c = \pm\sqrt{3} + 1$$

$c = \sqrt{3} + 1$. Reject $c = -\sqrt{3} + 1$ since it is not in the given interval.

2. $c = \pm\dfrac{\pi}{3}, \pm\dfrac{2\pi}{3}, 0$

$2\cos(-3\pi) = 2\cos(3\pi) = -2$

$f'(c) = 0 \rightarrow -6\sin(3c) = 0 \rightarrow \sin(3c) = 0 \rightarrow 3c = \pm\pi \rightarrow c = \pm\dfrac{\pi}{3}$.

Also, $3c = \pm 2\pi \rightarrow c = \pm\dfrac{2\pi}{3}$. And $3c = 0 \rightarrow c = 0$. Notice that

$3c = \pm 3\pi \rightarrow c = \pm\pi$ but these c values are the endpoints of the given interval and thus they must be rejected.

3. Yes, $c = e - 1$
   $y = \ln(x)$ does satisfy the Mean Value Theorem on $[1, e]$ because it is continuous on $[1, e]$ and differentiable on $(1, e)$.
   $$f'(c) = \frac{f(b) - f(a)}{b - a} \rightarrow \frac{1}{c} = \frac{f(e) - f(1)}{e - 1} \rightarrow \frac{1}{c} = \frac{1}{e - 1} \rightarrow c = e - 1.$$

4. No, $y = \ln(x)$ does not satisfy Rolle's Theorem on any interval because there are no x-values a and b such that $f(a) = f(b)$.

5. Because f is a polynomial, it continues and must obey the intermediate value theorem. There must be a root between $x = 1$ and $x = 2$. There is a root at $x = 3$ and there must be a root between $x = 4$ and $x = 5$.

6. Critical points: $(0, 2)$, $\left(\pm\sqrt{\dfrac{3}{2}}, \dfrac{-1}{4}\right)$; inflection points: $\left(\pm\dfrac{\sqrt{2}}{2}, \dfrac{3}{4}\right)$.

   Relative maximum: $(0, 2)$, Absolute minimum y value: $y = -\dfrac{1}{4}$.

Critical points:

$$y' = 4x^3 - 6x = 0 \rightarrow y' = 2x(2x^2 - 3) = 0 \rightarrow x = 0, \pm\sqrt{\frac{3}{2}}.$$

To find *y*-values, substitute *x*-values, into original function.

$$y = (0)^4 - 3(0)^2 + 2 = 2 \ (0, 2)$$

$$y = \left(\pm\sqrt{\frac{3}{2}}\right)^4 - 3\left(\pm\sqrt{\frac{3}{2}}\right)^2 + 2 = \frac{9}{4} - \frac{9}{2} + 2 = \frac{-1}{4} \quad \left(\pm\sqrt{\frac{3}{2}}, \frac{-1}{4}\right)$$

Inflection points: $y'' = 12x^2 - 6 = 0 \rightarrow x = \pm\frac{\sqrt{2}}{2}.$

$$\begin{array}{ccccccc} y'' & + & 0 & - & 0 & + \\ \hline x & \textcircled{-1} & -\frac{\sqrt{2}}{2} & \textcircled{0} & \frac{\sqrt{2}}{2} & \textcircled{1} \end{array}$$

Since there's a sign change at both $x = \frac{\sqrt{2}}{2}$ and $x = -\frac{\sqrt{2}}{2}$,

inflection points occur here. To find *y* values, substitute $x = \pm\frac{\sqrt{2}}{2}$

into original function: $y = \left(\pm\frac{\sqrt{2}}{2}\right)^4 - 3\left(\pm\frac{\sqrt{2}}{2}\right)^2 + 2 = \frac{1}{4} - \frac{3}{2} + 2 = \frac{3}{4}$

$\left(\pm\frac{\sqrt{2}}{2}, \frac{3}{4}\right)$. Absolute minimum value of *y* and relative maximum

points are found using the sign analysis chart for *y'*:

$$\begin{array}{ccccccccc} y & - & 0 & + & 0 & - & 0 & + \\ \hline x & \textcircled{-2} & -\frac{\sqrt{3}}{2} & \textcircled{-1} & 0 & \textcircled{1} & \frac{\sqrt{3}}{2} & \textcircled{2} \end{array}$$

Candidate for maximum: $x = 0$ because *y'* changes sign from
positive to negative. Substitute $x = 0$ into the original equation
to find y: $y = 0^4 - 3(0)^2 + 2 = 2$. Relative maximum point: (0, 2).

The absolute minimum occurs at $x = \pm\sqrt{\frac{3}{2}}$ because *y'* changes

sign from negative to positive. The *y*-value for these *x*-values

is $y = -\frac{1}{4}.$

7. Make a sign analysis chart for $f'(x)$ and $f''(x)$ based on the given graph.

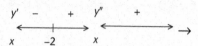

original function is decreasing on $(-\infty, -2)$ and increasing on $(-2, +\infty)$ and it's concave up.

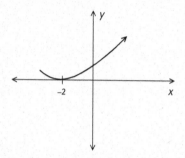

CHAPTER 7
SOLUTIONS

1. $A = 2$

$$A = \int_0^2 |1 - x^2|\,dx$$

$$A = \int_0^1 (1 - x^2)\,dx - \int_1^2 (1 - x^2)\,dx$$

$$A = x - \frac{x^3}{3}\Bigg]_0^1 - \left(x - \frac{x^3}{3}\right)\Bigg]_1^2$$

$$A = \left(1 - \frac{1}{3}\right) - \left[\left(2 - \frac{8}{3}\right) - \left(1 - \frac{1}{3}\right)\right]$$

$$= \frac{2}{3} - 1 + \frac{7}{3} = 2$$

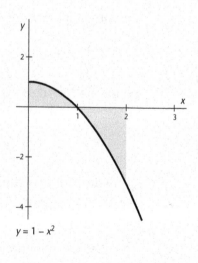

$y = 1 - x^2$

2. (A) 21 (B) –30

(A) $\int\limits_a^b [3f(x) - 2g(x)]dx = 3\int\limits_a^b f(x)dx - 2\int\limits_a^b g(x)dx = 3(5) - 2(-3) = 21$

(B) $\int\limits_b^a 6f(x)dx + \int\limits_a^a \dfrac{g(x)}{\pi}\,dx = -\int\limits_a^b 6f(x)dx + 0 = -6(5) = -30$

3. (A) $\sqrt{3x+1}$ (B) $\dfrac{8x}{e^{4x^2}}$

(A) $\dfrac{d}{dx}\int\limits_0^x \sqrt{3t+1}\,dt = \sqrt{3x+1}$

(B) $\dfrac{d}{dx}\int\limits_0^{4x^2} \dfrac{1}{e^t}\,dt = \dfrac{8x}{e^{4x^2}}$

4. $\dfrac{1}{e^4}$

$\int\limits_4^\infty e^{-x}dx = -e^{-x}\Big]_4^\infty \lim\limits_{\ell \to \infty}(-e^{-\ell} + e^{-4}) = e^{-4} = \dfrac{1}{e^4}$

5. (A) 8.382   (B) 9.832

(C) 9.143   (D) 9.107

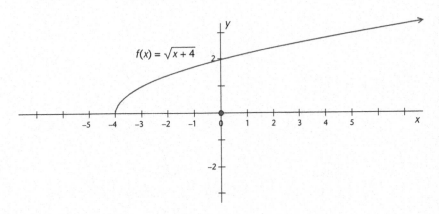

$f(x) = \sqrt{x+4}$

This is what the function should look like for parts (A) through (D) with their respective rectangles between $x = -3$ and $x = 2$.

(A)

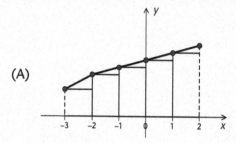

LRAM $f(x) = \sqrt{x + 4}$
$A \approx 1(f(-3) + f(-2) + f(-1) + f(0) + f(1)) = 8.382$.
This is an underestimation.

(B)

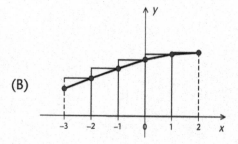

RRAM

$A \approx 1(f(2) + f(1) + f(0) + f(-1) + f(-2)) = 9.832$.
This is an overestimation.

(C)

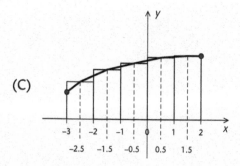

MRAM
$A \approx 1(f(-2.5) + f(-1.5) + f(-.5) + f(.5) + f(1.5)) = 9.143$.

(D)

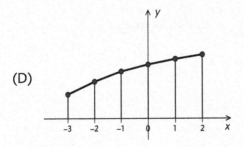

TRAP

$$A = \frac{1}{2} \cdot 1(f(-3) + 2f(-2) + 2f(-1) + 2f(0) + 2f(1) + f(2)] =$$

9.107.

6. $\frac{2}{3}e^{\frac{3x}{2}} + C$ *u*-substitution

$$\int e^x \sqrt{e^x}\, dx \quad u = e^x \quad du = e^x\, dx$$

$$\int \sqrt{u}\, du = \int u^{\frac{1}{2}}\, du = \frac{2}{3}u^{\frac{3}{2}} + C = \frac{2}{3}e^{\frac{3x}{2}} + C$$

7. $\frac{\sin^4(4x)}{16} + C$ *u*-substitution

$$\int \sin^3(4x) \cos(4x)\, dx \quad u = \sin(4x)$$

$$du = 4\cos(4x)\, dx \quad \frac{1}{4}du = \cos(4x)\, dx$$

$$\frac{1}{4}\int u^3\, du = \frac{1}{4}\frac{u^4}{4} + C = \frac{u^4}{16} + C = \frac{\sin^4(4x)}{16} + C$$

8. $\dfrac{x^2 \ln x}{2} - \dfrac{x^2}{4} + C$ Integration by parts

$$\int x \ln x \, dx \qquad\qquad u = \ln x \quad v = \dfrac{x^2}{2}$$

$$\dfrac{x^2}{2} \ln x - \dfrac{1}{2} \int x \, dx \qquad du = \dfrac{1}{x} dx \quad dv = x \, dx$$

$$\dfrac{x^2}{2} \ln x - \dfrac{x^2}{4} + C$$

9. $4 \ln|x - 2| - 2 \ln|x - 1| + C$ Integration by partial fractions

$$\int \dfrac{2x}{x^2 - 3x + 2} dx = \int \dfrac{2x}{(x - 2)(x - 1)} dx \qquad \dfrac{A}{x - 2} + \dfrac{B}{x - 1} =$$

$$\dfrac{2x}{(x - 2)(x - 1)} \;\rightarrow\; Ax - A + Bx - 2B = 2x \;\rightarrow\; A + B = 2 \quad \text{and}$$

$$-A - 2B = 0 \xrightarrow[\text{EQUATIONS}]{\text{ADD}} -B = 2 \rightarrow B = -2 \rightarrow A = 4$$

$$\int \dfrac{2x}{x^2 - 3x + 2} dx = \int \dfrac{4}{x - 2} dx + \int \dfrac{-2}{x - 1} dx = 4 \ln|x - 2| - 2 \ln|x - 1| + C$$

10. $\dfrac{2\sqrt{3}}{3} \tan^{-1}\left(\dfrac{x}{\sqrt{3}}\right) + C$ $u =$ substitution

$$\int \dfrac{2}{3 + x^2} dx = \dfrac{2}{3} \int \dfrac{1}{1 + \frac{x^2}{3}} dx = \dfrac{2}{3} \int \dfrac{1}{1 + \left(\frac{x}{\sqrt{3}}\right)^2} dx \quad u = \dfrac{x}{\sqrt{3}} \quad du = \dfrac{1}{\sqrt{3}} dx$$

$$\sqrt{3} \, du = dx$$

$$\dfrac{2\sqrt{3}}{3} \int \dfrac{1}{1 + u^2} du = \dfrac{2\sqrt{3}}{3} (\tan^{-1} u) + C = \dfrac{2\sqrt{3}}{3} \tan^{-1}\left(\dfrac{x}{\sqrt{3}}\right) + C$$

CHAPTER 8
SOLUTIONS

1. $\int y\,dx = \int x\,dx$

$$\frac{y^2}{2} = \frac{x^2}{2} + C$$

$$y^2 = x^2 + C \Rightarrow y = \pm\sqrt{x^2 + C}$$

2. $\int \frac{dy}{y} = \int \frac{dx}{x}$

$$\ln|y| = \ln|x| + C$$

$$y = e^{\ln|x|+C} = e^{\ln|x|} \cdot e^C = Cx$$

3.

(x, y)	−2	−1	0	1	2
2	−4	−3	−2	−1	0
1	−3	−2	−1	0	1
0	−2	−1	0	1	2
−1	−1	0	1	2	3
−2	0	1	2	3	4

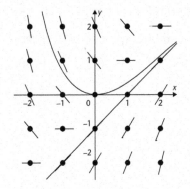

4. $\dfrac{dP}{dt} = kP \Rightarrow P = P_0 e^{kt}$

$50 = 100e^{5750k} \Rightarrow k = \dfrac{\ln 0.5}{5750} = -0.00012054$

$P = 100e^{5065k} = 54.304\%$

5. $y = 2e^{x^2} - 1$

$\dfrac{dy}{dx} = 2x(y+1) \rightarrow \dfrac{dy}{y+1} = 2x\,dx \rightarrow \ln|y+1| = x^2 + C \xrightarrow{(0,\,1)}$

$\ln 2 = C \rightarrow \ln|y+1| = x^2 + \ln 2 \rightarrow \ln\dfrac{|y+1|}{2} = x^2 \rightarrow \dfrac{|y+1|}{2} = e^{x^2} \rightarrow$

$|y+1| = 2e^{x^2} \rightarrow y + 1 = 2e^{x^2}$ OR $y + 1 = 2e^{x^2}$. Since (0, 1) satisfies

$y + 1 = 2e^x$, $y = 2e^{x^2} - 1$.

6. (A) When $P = 5$ (population is 5 million)   (B) 10 million

(A) Logistic growth equation $\dfrac{dP}{dt} = \dfrac{K}{M}P(M - P)$. Population grows

fastest when $P = \dfrac{M}{2}$. In $\dfrac{dP}{dt} = \dfrac{3}{10}P(10 - P)$, $P = \dfrac{10}{2} \rightarrow P = 5$.

(B) $\lim\limits_{t \to \infty} P(t) = \lim\limits_{t \to \infty} \dfrac{M}{1 + Ae^{-kt}} = M$. In this problem, $M = 10$. This means

that as time goes on, the population approaches 10 million.

CHAPTER 9
SOLUTIONS

1. $\dfrac{\left[\dfrac{2}{3}x^{3/2}\right]_1^9}{8} = \dfrac{18-\dfrac{2}{3}}{8} = \dfrac{54-2}{24} = \dfrac{13}{6}$

2. $f_{avg} = \dfrac{\displaystyle\int_0^2 (4-x^2)\,dx}{2-0} = \dfrac{\left[4x-\dfrac{x^3}{3}\right]_0^2}{2} = \dfrac{8-\dfrac{8}{3}}{2} = \dfrac{8}{3}$

   $f(c) = 4-c^2 = \dfrac{8}{3} \Rightarrow c^2 = \dfrac{4}{3} \Rightarrow c = \dfrac{2}{\sqrt{3}}$

3. Disp $= \displaystyle\int_0^5 (12-3t)\,dt = \left[12t-\dfrac{3t^2}{2}\right]_0^5 = 60-\dfrac{75}{2} = 22.5$ ft

   Dist $= \displaystyle\int_0^5 |12-3t|\,dt$

   Dist $= \displaystyle\int_0^4 (12-3t)\,dt - \int_4^5 (12-3t)\,dt$

   Dist $= \left[12t-\dfrac{3t^2}{2}\right]_0^4 - \left[12t-\dfrac{3t^2}{2}\right]_4^5$

   Dist $= 48-24-\left[60-\dfrac{75}{2}-(48-24)\right] = 25.5$ ft

4.

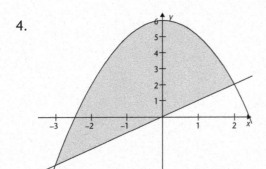

   $6-x^2 = x \Rightarrow (x+3)(x-2) = 0 \Rightarrow x = -3, x = 2$

   $A = \displaystyle\int_{-3}^2 (6-x^2-x)\,dx = \dfrac{125}{6}$

5. $12 + \int_0^{45} E(t) = 12 + 10(10) + \dfrac{1}{2}(10)(10) + \dfrac{1}{2}(15)(15) + \dfrac{10}{2}(15+5) \approx 375$

6. $L = \int_0^1 \sqrt{1 + e^{2t}}\,dt = 2.003$

7.

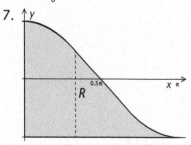

$$V = \pi \int_0^\pi \left(\cos x - (-1)\right)^2 dx = \pi \int_0^\pi \left(\cos x + 1\right)^2 dx$$

CHAPTER 10
SOLUTIONS

1. $\begin{cases} x(t) = 2\sin t \\ y(t) = 3\cos t \end{cases} \rightarrow \begin{cases} \dfrac{x}{2} = \sin t \\ \dfrac{y}{3} = \cos t \end{cases} \rightarrow \begin{cases} \dfrac{x^2}{4} = \sin^2 t \\ \dfrac{y^2}{9} = \cos^2 t \end{cases} \rightarrow$

$\dfrac{x^2}{4} + \dfrac{y^2}{9} = 1.$ Ellipse $a = 2$, $b = 3$.

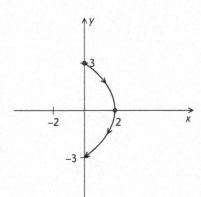

$$0 \le t \le \pi$$

t	0	$\pi/2$	π
x	0	2	0
y	3	0	-3

Don't forget to indicate direction of motion.

2. $r(t) = \left(\dfrac{t}{2}\right)i + (e^t)j \quad t > 0.$

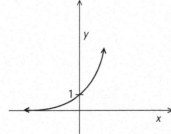

In parametric form, $\begin{cases} x(t) = \dfrac{t}{2} \\ y(t) = e^t \end{cases} \quad t > 0.$

In Cartesian form, $y = e^{2x}, \quad x > 0.$

3. $r = 2 - 3\cos\theta$ is a looped limaçon.

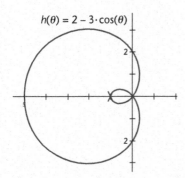

$h(\theta) = 2 - 3\cdot\cos(\theta)$

4. $r = 4\sin(2\theta)$ has 4 petals and the length of each petal is given by

$$L = \int_{0}^{\pi/2} \sqrt{16\sin^2(2\theta) + 64\cos^2(2\theta)}\, d\theta = 9.688$$

5. $r = \cos\theta \rightarrow r^2 = r\cos\theta \rightarrow x^2 + y^2 = x \rightarrow x^2 - x + y^2 = 0 \rightarrow$

$x^2 - x + \dfrac{1}{4} + y^2 = \dfrac{1}{4} \rightarrow \left(x - \dfrac{1}{2}\right)^2 + y^2 = \dfrac{1}{4}$. This is a circle with the

center at $\left(\dfrac{1}{2}, 0\right)$ and radius $\dfrac{1}{2}$.

6. $x = 2 \rightarrow r\cos\theta = 2 \rightarrow r = \dfrac{2}{\cos\theta} \rightarrow r = 2\sec\theta$.

7. $\dfrac{dy}{dx} = \dfrac{dy/dt}{dx/dt} = \dfrac{-2\sin t}{2\cos t} = -\tan t \Rightarrow \dfrac{dy}{dx}_{t=5\pi/4} = -1$

$\dfrac{d^2y}{dx^2} = \dfrac{\dfrac{d(dy/dt)}{dt}}{dx/dt} = \dfrac{-\sec^2 t}{2\cos t} = \dfrac{-\sec^3 t}{2} \Rightarrow \dfrac{d^2y}{dx^2}_{t=5\pi/4} = \dfrac{-\left(-\sqrt{2}\right)^3}{2} = \sqrt{2}$

8.

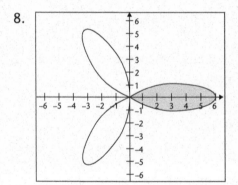

$A = 2\left(\dfrac{1}{2}\right)\displaystyle\int_0^{\pi/6} 36\cos^2 3\theta \, d\theta$

$A = 9.425$

9. $\mathbf{v}(t) = \dfrac{3}{2}\cos\dfrac{t}{2}\mathbf{i} - \dfrac{3}{2}\sin\dfrac{t}{2}\mathbf{j}$

$\mathbf{a}(t) = \dfrac{-3}{4}\sin\dfrac{t}{2}\mathbf{i} - \dfrac{3}{4}\cos\dfrac{t}{2}\mathbf{j}$

Speed $= \sqrt{\dfrac{9}{4}\cos^2\dfrac{t}{2} + \dfrac{9}{4}\sin^2\dfrac{t}{2}} = \dfrac{3}{2}$

Speed is a constant.

CHAPTER 11
SOLUTIONS

1. 0

$$\lim_{n\to\infty}\left\{\frac{2n-1}{3en^2-1}\right\}=\lim_{n\to\infty}\left\{\frac{2n}{3en^2}\right\}=0$$

2. (A) converges   Direct Comparison test with $\sum\limits_{k=1}^{\infty}\frac{\cos k}{k^2}$

   (B) diverges   Divergence test

   (C) converges   Ratio test or alternating test

(A) $\sum\limits_{k=1}^{\infty}\frac{\cos k}{k^2}$ $\sum\limits_{k=1}^{\infty}\left|\frac{\cos k}{k^2}\right|=\sum\limits_{k=1}^{\infty}\frac{|\cos k|}{k^2}\le\sum\limits_{k=1}^{\infty}\frac{1}{k^2}$ convergent

p series. By the comparison test $\sum\limits_{k=1}^{\infty}\left|\frac{\cos k}{k^2}\right|$ converges

therefore $\sum\limits_{k=1}^{\infty}\frac{\cos k}{k^2}$ converges. Remember, if a series

converges absolutely, then it converges.

(B) $\sum\limits_{k=1}^{\infty}\frac{e^k}{k+1}$. Since $\lim\limits_{K\to\infty}\frac{e^k}{k+1}\ne 0$, series diverges.

(C) Converges   Ratio test for absolute convergence

$\sum\limits_{k=1}^{\infty}\frac{(-1)^k}{(k-1)!}$ $\lim\limits_{k\to\infty}\left|\frac{(-1)^{k+1}}{(k+1-1)!}\circ\frac{(k-1)!}{(-1)^k}\right|=\lim\limits_{k\to\infty}\frac{(k-1)!}{k!}$

$$=\lim_{k\to\infty}\frac{1}{k}=0$$

This is an alternating series whose terms are getting smaller. Therefore it converges.

3. A Maclaurin polynomial is a Taylor polynomial centered at $x=0$.

4. Interval of convergence $1 \le x \le 3$, radius of convergence: 1

$$\sum_{k=1}^{\infty} \frac{(2-x)^k}{k^3} \qquad \lim_{k \to \infty} \left| \frac{(2-x)^{k+1}}{(k+1)^3} \cdot \frac{k^3}{(2-x)^k} \right| = \lim_{k \to \infty} \left| (2-x) \frac{k^3}{(K+1)^3} \right| =$$

$$\lim_{k \to \infty} \left| 2-x \right| = \left| 2-x \right| < 1 \to -1 < 2-x < 1 \to 1 < x < 3$$

Check endpoints! At $x = 1$ $\displaystyle\sum_{k=1}^{\infty} \frac{1}{k^3}$ convergent p series. At $x = 3$,

$\displaystyle\sum_{k=1}^{\infty} \frac{(-1)^k}{k^3}$ converges absolutely so it converges. Interval of

convergence is $1 \le x \le 3$. Radius is 1.

5. $P_4(x) = 3e + 3e(x-1) + \dfrac{3e}{2}(x-1)^2 + \dfrac{3e}{6}(x-1)^3 + \dfrac{3e}{24}(x-1)^4$

$P_4(x) = f(a) + f'(a)(x-a) + \dfrac{f''(a)(x-a)^3}{2!} + \dfrac{f'''(a)(x-a)^3}{3!} +$

$\dfrac{f^{IV}(a)(x-a)^4}{4!}$

$a = 1$, $f(x) = 3e^x$ $\qquad f'(x) = f''(x) = f'''(x) = f^{IV}(x)$
$f(a) = f(1) = 3e$
$f'(a) = f'(1) = 3e$
$f''(1) = f'''(1) = f^{IV}(1) = 3e$

$P_4(x) = 3e + 3e(x-1) + \dfrac{3e}{2}(x-1)^2 + \dfrac{3e}{6}(x-1)^3 + \dfrac{3e}{24}(x-1)^4$

6. $\dfrac{\sin(2x)}{x} = 2x - \dfrac{2^3 x^2}{3!} + \dfrac{2^5 x^4}{5!} - \dfrac{2^7 x^6}{7!} + \cdots - \infty < x < \infty$

Maclaurin series for $y = \sin x$ is

$$\sin x = x - \frac{x^3}{3!} + \frac{x^5}{5!} - \frac{x^7}{7!} \quad -\infty < x < \infty$$

$$\sin(2x) = 2x - \frac{(2x)^3}{3!} + \frac{(2x)^5}{5!} - \frac{(2x)^7}{7!} \quad -\infty < x < \infty$$

$$\frac{\sin 2x}{x} = 2 - \frac{1}{x}\frac{(2x)^3}{3!} + \frac{1}{x}\frac{(2x)^5}{5!} - \frac{1}{x}\frac{(2x)^7}{7!} \quad -\infty < x < \infty$$

$$\frac{\sin 2x}{x} = 2 - \frac{2^3 x^2}{3!} + \frac{2^5 x^4}{5!} - \frac{2^7 x^6}{7!} \quad -\infty < x < \infty$$

Notes

Notes

Notes

Notes

Notes

Notes

Notes